T0135572

Dynamics of the IL-2 cytokine network and T-cell proliferation

DISSERTATION

zur Erlangung des akademischen Grades
doctor rerum naturalium
(Dr. rer. nat.)
im Fach Biophysik

eingereicht an der
Mathematisch-Naturwissenschaftlichen Fakultät I
Humboldt-Universität zu Berlin

von
Frau Dipl.-Biophys. Dorothea Busse
geboren am 23.01.1975 in Göttingen

Präsident der Humboldt-Universität zu Berlin:
Prof. Dr. Dr. h.c. Christoph Markschies

Dekan der Mathematisch-Naturwissenschaftlichen Fakultät I:
Prof. Dr. Lutz-Helmut Schön

Gutachter:

1. Prof. Thomas Höfer, PhD
2. Dr. habil. Alexander Scheffold
3. Dr. Michal Or-Guil

eingereicht am: 17.08.2009
Tag der mündlichen Prüfung: 14.12.2009

Bibliografische Information der Deutschen Nationalbibliothek

Die Deutsche Nationalbibliothek verzeichnet diese Publikation in der
Deutschen Nationalbibliografie; detaillierte bibliografische Daten sind
im Internet über http://dnb.d-nb.de abrufbar.

ISBN 978-3-8325-2693-1

Logos Verlag Berlin GmbH
Comeniushof, Gubener Str. 47,
10243 Berlin
Tel.: +49 (0)30 42 85 10 90
Fax: +49 (0)30 42 85 10 92
INTERNET: http://www.logos-verlag.de

Zusammenfassung

In dieser Arbeit wird die Kommunikation von T-Helfer-Lymphozyten und regulatorischen T-Lymphozyten mittels des Zytokins Interleukin-2 untersucht. T-Helfer-Lymphozyten (Th-Zellen) spielen eine zentrale Rolle bei adaptiven Immunantworten, wohingegen regulatorische T Zellen (Treg-Zellen) die Toleranz des Immunsystems gegenüber Selbstantigenen aufrecht erhalten. Treg-Zellen verhindern die Proliferation von autoreaktiven Th-Zellen und können wahrscheinlich auch die Vermehrung antigenspezifischer T-Zellen bei Immunantworten und damit entzündliche Immunpathologien begrenzen. Beide Zelltypen benötigen Interleukin-2 (IL-2) um ihre Aufgabe im Immunsystem wahrnehmen zu können. IL-2 wird überwiegend von antigenstimulierten Th-Zellen sezerniert, denen es als autokriner Wachstumsfaktor dient. Treg-Zellen können selbst kein IL-2 produzieren, bedürfen jedoch dieses Zytokins für ihre Entwicklung und Proliferation. IL-2-Signale werden über hochaffine heterotrimere IL-2-Rezeptoren vermittelt, die konstitutiv von Treg-Zellen (nicht jedoch von Th-Zellen) exprimiert werden. Deshalb wurde vermutet, dass Treg-Zellen antigenstimulierten Th-Zellen einen Wachstumsfaktor entziehen und sie dadurch an der Teilung hindern können. Experimentelle Studien *in vitro* und *in vivo* konnten diese Hypothese bestätigen. Allerdings wird der hochaffine IL-2-Rezeptor auch in antigenstimulierten Th-Zellen induziert, und die Rezeptorexpression wird in Th- und Treg-Zellen durch IL-2 verstärkt. Durch diese vielfältigen Zellinteraktionen und Rückkopplungen entsteht ein komplexes Regulationsnetzwerk, dessen Dynamik bisher nur unzureichend verstanden ist.

Die Analyse der IL-2-Netzwerkdynamik und insbesondere der Konkurrenz um IL-2 durch Th- und Treg-Zellen ist Ziel dieser Arbeit. Hierzu wurde eine kombinierte Analyse des Systems mittels eines mathematischen Modells und Zellkulturexperimenten durchgeführt. In den Experimenten wurden Th-Zellen allein oder in Anwesenheit mit Treg-Zellen kultiviert und ihr IL-2 Rezeptorexpressionslevel für verschiedene Antigenstimulusstärken untersucht. Als mathematisches Modell wurde ein Reaktions-Diffusions-Modell

1

entwickelt, das die Dynamik sowohl der IL-2 Rezeptoren auf einer Th-Zelle als auch auf einer Treg-Zellen beschreibt, sowie die Dynamik des extrazellulären IL-2. Näherungsweise haben wir eine räumlich eindimensionale Formulierung betrachtet, die die Kommunikation von benachbarten Zellen durch IL-2 beschreibt. Als weitere Vereinfachung konnten wir annehmen, dass gemessen an der Zeitskale der IL-2 und IL-2-Rezeptor-Expression (Stunden) sich sehr schnell (Sekunden) ein quasistationäres Profil der extrazellulären IL-2-Konzentration einstellt. Diese Annahmen ermöglichten es uns, das Model auf ein System aus gewöhnlichen Differenzialgleichungen zu reduzieren und eine detaillierte Analyse des dynamischen Verhaltens und seiner Parameterabhängigkeit durchzuführen.

Mit diesem Modell wurde in Kapitel 3 zuerst die Dynamik der IL-2 Rezeptor-Expression einer einzelnen Th-Zellen untersucht. Die Expression der IL-2 Rezeptoren wird reguliert durch eine autokrine positive Rückkopplung des Signals auf seine eigene Expression. Des Weiteren zeichnet sich die Dynamik durch stimulationsabhängige Internalisierung des Rezeptor und seines Recyclings aus. Besonders die Regulation der IL-2 Rezeptorproduktion durch die autokrine positive Rückkopplung bestimmt die Dynamik des Rezeptors. Auf Grund dieser Regulation erfolgt die Expression des IL-2 Rezeptors als ein Alles-oder-Nichts Ereignis. Hohe Expression der IL-2 Rezeptoren wird auch als Aktivierung bezeichnet. Zusammen mit der stimulusinduzierten Internalisierung des Rezeptors und des Recyclings führt sie dazu, dass IL-2 im hohen Maße von den sezernierenden Zellen wieder aufgenommen wird, wenn diese viele Rezeptoren auf ihrer Oberfläche exprimieren. Diese wichtige Rolle der autokrinen positiven Rückkopplungsschleife für die Regulation der IL-2 Rezeptor Expression wurde durch Zellkulturexperimente bestätigt.

In Kapitel 4 wurde die Kommunikation zwischen Th-Zellen und Treg-Zellen mit Hilfe des oben beschriebenen Modells analysiert. Auch die Dynamik der IL-2 Rezeptoren auf Treg-Zellen ist bestimmt durch positive Rückkopplung. Interessanterweise bewirkt die positive Rückkopplung auf Treg-Zellen einen kontinuierlichen Anstieg in der Zahl der IL-2 Rezeptoren, da diese schon konstitutiv den IL-2 Rezeptor exprimieren. Dieser Unterschied in der IL-2 Rezeptor Regulation konnte experimentell bestätigt werden. Für die Konkurrenz um IL-2 ist diese unterschiedliche Regulation der IL-2 Rezeptor Expression von großer Bedeutung. Die Treg-Zellen können ihre Anzahl an IL-2 Rezeptoren dem extrazellularen IL-2 anpassen, während für die Th-Zellen eine kritische extrazelluläre IL-2 vorhanden sein muss, um die Anzahl der Rezeptoren zu erhöhen, dieses erfolgt dann sprunghaft. Die Möglichkeit, die Anzahl der IL-2 Rezeptoren der extrazellulären IL-2 Konzentration anzupassen, zusammen mit dem Vorteil, bereits eine hohe Anzahl an Rezeptoren zu exprimieren, macht Treg-Zellen zu effizienten Konkurrenten. Als Bedingungen für Konkurrenz konnten wir limitierte IL-2-Sekretion (z.b. durch moderat stimulierte Th-Zellen) und räumliche Nähe zwischen Treg- und Th-Zellen identifizieren. Abhängig von der Stärke des Antigen-Stimulus der Th-Zellen und der interzellulären Distanz können somit drei Möglichkeiten auftreten: (1) Proximale Treg-Zellen nehmen das gesamte verfügbare IL-2 auf und unterbinden vollständig autokrines IL-2-Signaling der Th-Zellen. (2) Th-Zellen können ungestört autokrines IL-2-Signaling etablieren, wenn Treg-Zellen sich zu weit entfernt befinden. (3) Beide Zelltypen werden aktiviert, wenn die IL-2-Sekretion durch aktivierte Th-Zellen sehr stark ist. Als ein weiteres Ergebniss dieser Arbeit konnte gezeigt werden, dass sowohl Konkurrenz als auch parakrine Aktivierung zwischen Th-Zellen möglich sind. Als Ausblick stellen wir einen mathematischen Ansatz vor, der es erlaubt, die Proliferationsparameter der Th-Zellen aus experimentellen Messungen mit dem Proliferationsfarbstoff CFSE zu bestimmen, Unsere ersten Ergebnisse von Zellkulturexperimenten legen nahe, dass Treg-Zellen eine reduzierte Teilung der Th-Zellen und eine beschleunigte Apoptose der sich teilenden Th-Zellen bewirken.

4

Contents

Chapter 1

Introduction

The immune system of higher vertebrates is composed of a great variety of highly specialized cells. Molecular interactions can occur throughout the organism as these cells circulate through the blood and the lymph system. However, the peripheral lymphoid tissues, where immune responses are initiated, provide a special, highly structured place for them. In general, cellular communication in the immune system is mediated by receptor ligand interactions. If both receptor and ligand are membrane bound, the interplay is considered as cell-cell contact dependent. A prominent example is the activation of the T cell receptor by antigen. An other common form of communication is mediated by cytokines like interleukin 2. Cytokines are small soluble messenger molecules that regulate various activities of mammalian cell function. They can act on their own or in concert with other mechanisms dependent on cytokines or cell-cell contact. As an outcome of cytokine stimulation often gene-expression networks are activated [71] resulting in proliferation, differentiation or survival of these cells.

In this work, the interaction of T helper lymphocytes and regulatory T lymphocytes is studied. T helper lymphocytes, or Th cells, are involved in the activation of cell-mediated immune responses, whereas regulatory T cells (Treg cells) maintain tolerance to self-antigen. Th cells must be stimulated by specialized antigen presenting cells (APCs) for activation. Activated Th cells then differentiate into Th1, Th2 or Th17 cells depending on additional differentiation signals to mediate their effector function. This differentiation process also enables them to memorize their effector function for a faster and more efficient elimination of the pathogen upon restimulation. Autoimmune reactions are caused by an immune response against self-antigens. Treg cells prevent the activation and proliferation of autoreactive Th cells. They are therefore involved in the maintenance of tolerance. Treg cells also depend on antigen stimulation via an APC in order to exert their suppressive influence. How autoreactive Th cells and regulatory T cells interact is still under investigation. Cell-cell contact dependent mechanisms as well as cytokine-mediated communication are previously described [47, 53, 17, 6, 68, 70, 81, 82]. In this work, their interaction via the cytokine interleukin 2 (IL-2) is analyzed in detail. Both T lymphocyte types depend on IL-2; Th cells secrete IL-2 themselves and utilize it in an autocrine manner whereas Treg cells depend on paracrine IL-2.

Interleukin 2 belongs to the family of cytokines, that functions as messenger molecules primarily in the interaction of leukocytes. The action of cytokines in the immune system includes the control of cell proliferation, regulation of the immune response and hematopoesis. Typically, cytokines act on different cell types and therefore mediate diverse biological effects [1]. In this sense interleukin 2 is a typical member of its family. On the one hand it is involved in the activation and proper ongoing of an inflammatory response by acting on Th cells; on the other hand it is important for the maintenance of tolerance by preventing responses against self antigens by acting on Treg cells. Although the pleiotropic effect of cytokines is well excepted, the finding that IL-2 can promote immune responses or suppress them if required has been surprising. Competition for IL-2 has been proposed as a mechanism to control autocrine or paracrine action of IL-2 on Th cells or Treg cells, respectively, and therefore mediate immunostimulatory or immunosuppressive responses [17]. In this chapter IL-2 signaling in Th and Treg cells is introduced. The following overview about the complexity of IL-2 function and the resulting difficulty to define it emphasizes the importance to understand how IL-2 action is regulated.

1.1 IL-2 signaling in T helper lymphocytes

IL-2 is predominantly expressed when naive or resting Th cells are stimulated by their cognate antigen through the T cell receptor (TCR). Moreover, it has been shown that dendritic cells also produce IL-2 after exposure to bacteria [33]. Secreted IL-2 acts in an autocrine manner, since Th cells also express the high affinity IL-2 receptor after antigenic stimulation. IL-2 also acts in a paracrine manner on other Th cells, which express the IL-2 receptor, or on Treg cells that constitutively express this receptor.

Induction of IL-2 gene-expression in T helper lymphocytes

Stimulation of the T cell receptor (TCR) by its cognate antigen (antibody generating molecule) activates several signal transduction pathways including MAP-kinase pathway (mitogen-activated protein kinase-pathway) and a Ca^{2+} dependent pathway leading to the activation of mainly three families of transcription factors, namely NFAT (nuclear factor of activated T-cells), AP-1 (activator protein 1) and NF-κB (nuclear factor kappa-light-chain-enhancer of activated B cells). Analysis of the predicted transcription factor binding sites reveals that the promoter region of the IL-2 gene consists almost exclusively of binding sites for these transcription factors [41, 31]. However, the strength of the antigen stimulus is reflected in the duration of IL-2 expression and the percentage of IL-2 expressing cells rather than in the amount of IL-2 secretion [60, 43]. Upon TCR activation, signaling via the co-stimulatory molecule CD28 is important for IL-2 gene transcription. CD28 signaling strongly enhances IL-2 transcription and stabilizes IL-2 mRNA [47].

Therefore, IL-2 gene-expression is mostly controlled by the T cell receptor and its co-stimulatory receptors. IL-2 is secreted to the extracellular medium, where it acts in an autocrine or paracrine manner.

The IL-2 signaling in T helper lymphocytes

IL-2 signaling is mediated by the specific IL-2 receptor. The IL-2 receptor consists of three subunits, the α-chain, the β-chain and the-$\gamma_c$chain. Only the α-chain is an exclusive component of the IL-2 receptor. The β-chain is shared with the IL-15 receptor and the γ_c-chain with IL-15, IL-7, IL-4 and IL-9 receptors [51]. All three subunits are required for a high binding affinity for IL-2 ($K_D = 10$ pM). The β- and γ_c-chains are responsible for the signal transduction and the α-chain increases the binding affinity by 100-fold. Naive and resting Th cells are negative for the IL-2R α-chain and upregulate it after antigen stimulation, whereas the β- and γ_c-chain are constitutively expressed.

In *ex vivo* isolated Th and Treg cells, the binding capacity of IL-2 strongly correlates with the expression of the IL-2R α-chain at 24 hours after TCR stimulation [19]. This finding suggests that expression of the α-chain also controls the signal strength downstream of the IL-2 receptor. Due to its dependence on antigenic stimulation the expression of the IL-2R α-chain is also used as a marker for Th cell activation.

The main signaling molecules activated downstream of the IL-2 receptor are STAT5 (signal transducer and activator of transcription 5) and SHC (src homologous and collagen protein). STAT5 acts directly as a transcription factor, whereas SHC activates the MAP-kinase pathway and the PI3K dependent pathways [47, 27]. For the STAT5 signaling pathway crosstalk exists with signaling pathway of other cytokines, which share the same receptor subunits. The MAP-kinase pathway and the PI3K (phosphoinositide 3-kinase)-dependent pathways are also activated by the TCR and the costimulus CD28. Therefore clear assignment of IL-2 receptor target genes has been difficult. In the early immune response upregulation of genes involved in the cell cycle, apoptosis inhibition, induction of cytokine and cytokine receptor expression and of the general transcriptional machinery are observed [40]. However, one clear target gene of IL-2 signaling is the IL-2R α-chain [75]. Several STAT5 binding sites are reported in the promoter of the IL-2R α-chain, same of them acting in a cooperative manner [44, 51, 57]. In addition, IL-2 signaling enhances the accessibility of the promotor region of the IL-2R α-chain locus by chromatin remodeling [64]. Thus, the IL-2R α-chain expression is strongly controlled by IL-2 signaling due to positive feedback regulation.

The observation that addition of blocking antibodies specific for IL-2 does not only inhibit IL-2R α-chain expression, but also decreases the number of IL-2 expressing cells [17] reveals autocrine positive feedback regulation also on IL-2 itself. Considering the connection of IL-2 and IL-2R α-chain gene expression via the autocrine positive feedback loop, a gene-expression network arises. It is initiated by the antigen stimulus via the TCR and its activation results in a high expression of the IL-2R α-chain. High IL-2R α-chain expression is associated with Th activation, since IL-2 receptor signaling induces various genes involved in proliferation, survival and differentiation.

IL-2 expression kinetics

The expression of cytokines is typically transient. The maximum of IL-2 secreting cells is observed around 20 hours after stimulation and cells stop to secret IL-2 after about 72 hours (Figure 1.1, thick black curves). Interestingly, at the maximum only 25% of the Th cell population express IL-2 in both experimental set ups. The percentage of secreting cells and the duration of IL-2 secretion strongly depend on the strength of antigen stimulation [60, 77]. As reported previously the percentage of IL-2 secreting cells also depends on the autocrine positive feedback induced by IL-2 signaling, since the number of IL-2 secreting cells is strongly reduced in the presence of blocking antibodies for IL-2 (Figure 1.1, left panel, black curve). Optimal Th cell proliferation and differentiation in cell culture are achieved by sequential

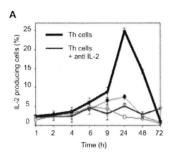 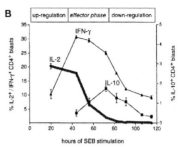

Figure 1.1. IL-2 secretion kinetics (A) *Ex vivo* isolated CD4$^+$ CD25$^-$ Th cells from DO11.10 mice were stimulated with 1µg/ml anti-CD3 and irradiated APCs. The percentage of IL-2 secreting cells (thick black curve) were determined by IL-2 secretion assay. (taken from [17]). (B) Naive Th cells were isolated from Balb/c mice and stimulated with SEB (staphylococcus aureus enterotoxin B) and APCs. IL-2 producing cells (thick black curve) was determined by intracellular staining after fixation (taken from [5]). The percentage of IL-2 secretion is transient, after approximately 20 h the maximum of about 25% percent of the Th cells secret IL-2. The secretion is downregulated after approximately 72 h.

stimulation of TCR and the IL-2 receptor [43], revealing that both signals are sufficient for Th cell activation. Limited antigen stimulus, negative feedback regulation by STAT5 among other cytokine receptor signal transduction molecules or even TCR-mediated inhibition are described to terminate IL-2 expression and secretion [77, 85, 48]. The relative importance of these mechanisms in the course of an immune response remains to be elucidated. In this work, predominantly, the activation of the IL-2/IL-2 receptor gene-expression network is studied.

1.2 T cell receptor and IL-2 receptor signaling in regulatory T cells

Most specific properties of Treg cells are mediated by the transcription factor FOXP3 (forkhead box P3) that is constitutively expressed in Treg cells but not in Th cells. However, the induction of FOXP3 in Th cells is sufficient to induce suppressive activity in these cells [38]. FOXP3 interacts with the transcription factor NFAT [89], which is activated by TCR stimulation. FOXP3 competes with and displaces AP-1, another transcription factor activated by

TCR signaling, in order to form a complex with NFAT. The NFAT/FOXP3 complex represses the transcription of the *Il2*-gene instead of inducing it as the NFAT/AP-1 complex does. In addition, an interaction with NF-ϰB, the third main transcription factor of antigen stimulation, has been reported [8].

Moreover, the interaction of FOXP3 and NFAT leads to an upregulation of the *Cd25* (IL-2R α-chain), *Ctla4* (cytotoxic T cell associated antigen-4) and *Gitr* (glucocorticoid-induced TNF receptor family-related protein) genes. These genes mediate or enhance the suppressive activity of regulatory T cells [89, 68]. Therefore, instead of initiating a gene-expression program that promotes an immune response, FOXP3 transforms the signaling via the TCR in such a way that it suppresses the relevant genes and activates gene-expression that is important for its suppressive function.

Although IL-2 expression is suppressed in regulatory T cells, paracrine IL-2 signaling is important for Treg cell function [30]. Treg cells constitutively express high amounts of the high-affinity IL-2 receptor due to the presence of the transcription factor FOXP3 in Treg cells. Thus Treg cells are responsive to IL-2 without antigen stimulation. Impaired IL-2 signaling also reduces the level of IL-2R α-chain expression in Treg cells [29]. Therefore the positive feedback of IL-2R α-chain signaling described in Th cells also exists in regulatory T cells. Moreover, IL-2 signaling is required for the maintenance of high expression levels of FOXP3 in the periphery [29]. Hence, high IL-2 receptor expression correlates with high FOXP3 expression level and subsequently strong suppressive activity. An additional role of IL-2 in regulatory T cell development and homeostasis was reported [29, 72, 11].

Surprising similarities exist between the signaling leading to an induction of an immune response by Th cells and the suppression of autoimmune reactions by Treg cells (Figure 1.2). Both involve TCR and IL-2 receptor signaling and positive feedback regulation at least of the IL-2R α-chain. The differences are mediated by the transcription factor FOXP3. In summary, TCR induces IL-2 and IL-2R α-chain gene-expression in Th cells that enables further autocrine positive feedback activation of both genes. The expression of the transcription factor FOXP3 leads to the suppression of IL-2 transcription and expression of other effector cytokines and to the constitutive expression of the IL-2R α-chain in Treg cells. IL-2 signaling enhances FOXP3 expression that leads to further upregulation of the IL-2R α-chain and mediates their suppressive capacity. In the following section the physiological role of IL-2 is introduced.

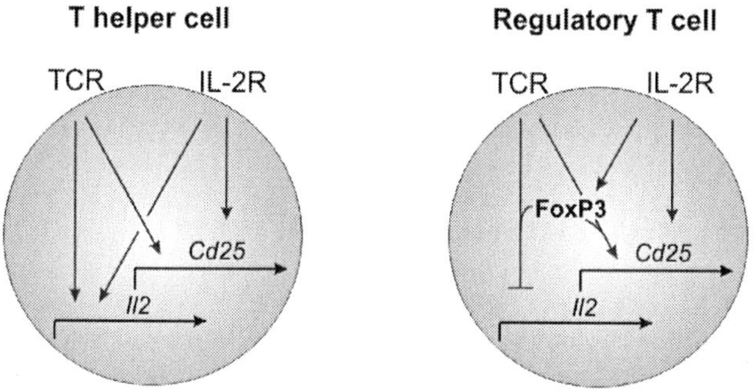

Figure 1.2. IL-2/IL-2R gene-expression network in Th and Treg cells
Expression of the *Il2* and the *Cd25* genes depends on the T cell receptor and the IL-2 receptor signaling. In Th cells, signaling of both receptors enhances the expression of these genes (left panel), whereas in Treg cells, IL-2 transcription is suppressed and constitutive expression of CD25 is enhanced due to the presence of the transcription factor FOXP3 (right panel).

1.3 Role of interleukin 2 in immunity and tolerance

Interleukin 2 was first described in the late 1970s as the essential factor for long-term culturing of T lymphocytes and therefore named T-cell growth factor [56, 55]. Its sole role as T cell growth factor was questioned when mice deficient for IL-2 or subunits of the IL-2 receptor were found to develop not the expected lymphopenic phenotype. Instead autoimmune diseases caused by hyper-proliferation and increased numbers of activated T and B cells were observed [66, 87, 79]. However, adoptive transfer studies about the role of IL-2 in mice with a normal immune system reveal the importance of IL-2 for a balanced immune response [18, 20, 88, 49].

Immunostimulatory function of IL-2

Detailed analysis of the role of IL-2 as a T cell growth factor in cell culture reveals that the initiation of proliferation in $CD4^+$ T cells is antigen dependent, whereas after 1-2 days T cell expansion and differentiation depend on IL-2 [43, 18]. Adoptive transfer of Th cells deficient for IL-2 or subunits of the IL-2 receptor into wild type mice enables studies about the function of IL-2 in a balanced immune system. In these experiments cells cultured previously or taken from other mice are transferred into a recipient mouse to study their behavior in the immunological environment typical for the recipient mouse. These studies confirmed that the initiation of Th cell proliferation depends on antigen stimulation rather than on IL-2. Nevertheless, transferred IL-2$^{-/-}$ cells failed to survive and to develop into effector Th cells. Consequently, the development into long-lived memory cells was impaired [18, 15]. In contrast, IL-2$^{-/-}$ cells cultured in the presence of IL-2 prior to their transfer survived over a long time [18]. In agreement with these observations, Th cells deficient for the IL-2R α-chain also showed reduced proliferation after transfer [49]. Therefore optimal T cell expansion, proper effector and memory function and long term survival depend on IL-2 in cell culture and in mice that do not show a lymphopenic phenotype or suffer from serious autoimmune diseases [39]. Dooms at el. [18] summarized these findings in concluding that IL-2 signaling mediates competitive fitness in CD4$^+$ T lymphocytes. Adoptive transfer studies on the role of IL-2 in CD8$^+$ T cell survival and function support this result [21, 20, 88]. However, it is still under investigation, if competitive fitness solely depends on IL-2 or if it can be replaced by cytokines that share parts of the receptor subunits with the IL-2 receptor, like IL-7 or IL-15.

In addition, IL-2 not only promotes immune responses, but also attenuates them. At high concentrations, IL-2 also enhances activation-induced cell death (AICD) [63]. As an overall picture, IL-2 is important for the balance of an immune response.

IL-2 exerts its immunosuppressive effect mainly by acting on regulatory T cells

The anti-inflammatory effect of IL-2 is revealed clearly by the observation that IL-2 and IL-2 receptor deficient mice develop serious autoimmune diseases. Mice deficient for IL-2 or for the α- or β-chain of the IL-2 receptor develop autoimmune diseases, like inflammatory bowel disease and hemolytic anemia. These effects are caused by hyperproliferation and increased activation of T and B lymphocytes in IL-2$^{-/-}$ and IL-2R α-chain$^{-/-}$ mice and by spontaneous T lymphocyte activation in IL-2 receptor β-chain$^{-/-}$ mice [66, 65, 79, 87]. For the IL-2R β-chain deficient mice it was shown that the diseases can be cured by bone marrow transplantation of wild type CD4$^+$ T cells [80]. More specifically, the subpopulation of CD25$^+$ cells of the transferred CD4$^+$ population is necessary to cure these autoimmune diseases [67]. This subpopulation was identified as CD4$^+$CD25$^+$ regulatory T cells. Furthermore, Furtado et al. showed that impaired IL-2 signaling in regulatory T cells causes the autoimmune reactions [30]. Therefore IL-2 signaling in regu-

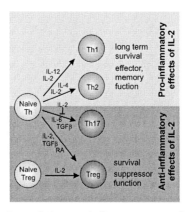

Figure 1.3. Overview of the pro-inflammatory and anti-inflammatory effects of IL-2 The pro-inflammatory effects of IL-2 involve long term survival and memory function of Th1 and Th2 cells. The anti-inflammatory effects are the inhibition of Th17 cells and the conversion of naive Th cells into induced Treg cells. IL-2 enhances the activation, survival and suppressor function of natural occurring Treg cells (adapted from [39]).

latory T cells is important to maintain tolerance. In addition, regulatory T cells are involved in the control of the immune response to microbial infections [7].

Overview of pro- and anti-inflammatory effects of IL-2

To complete the picture of the pro- and anti-inflammatory effects of IL-2 an overview is given (Figure 1.3): (1) The pro-inflammatory action on Th cell activation, long term survival of the effector cells and their memory function were discussed in the previous section. (2) Moreover, IL-2 inhibits the differentiation of IL-17 producing Th17 cells. IL-17 has been identified as an important cytokine in promoting allergy and tissue-specific autoimmune diseases. IL-2 inhibits the expression of IL-6, which is together with IL-1 and TGF-β crucial for Th17 induction. Therefore, IL-2 can inhibit the induction of Th17 cells [47] and attenuate inflammation. (3) IL-2 is important in the maintenance of tolerance by inducing the conversion of naive Th cells into inducible Treg cells in the presence of TGF-β and retinoic acid (RA). However, the role of these inducible Tregs and their stability still need to be determined [68]. (4) In natural occurring Treg cells that are derived from the thymus, IL-2 signaling promotes their activation, survival and suppressor

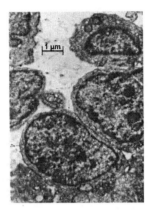

Figure 1.4. Electron microscopical picture of lymphoid cells The intercellular distance between lymphoid cells is less than a few micrometers (spleen of
a rabbit, taken from [14]).

function as presented above. In the following, the term Treg cells always
refers to thymus-derived Treg cells. Thus IL-2 mediates several pro-and anti-
inflammatory effects.

1.4   Space as a parameter in T helper and regulatory T cell interaction

Before analyzing the importance of the spatial organization of Th and Treg
cell interaction, it is instructive to present what is known about the localization of Th and Treg cells *in vivo*. Immune responses are initiated by the
APC/Th cell interaction in the secondary lymphoid tissues, such as lymph
nodes. Activated Th cells differentiate into effector Th cells, before they migrate to the site of infection. Treg cells are also stimulated by antigen in the
peripheral lymphoid tissues and migrate to the site of infection [68, 1]. Therefore, the Th/Treg cell interaction can take place in the lymphoid tissues as
well as at the site of infection. However, the initiation of an immune response
takes place in the lymphoid tissues. Lymphoid organs such as lymph nodes
and spleen are highly structured. Within both organs there are areas where
mostly T cells are located, which is the paracortical area in a lymph node
and the periarteriolar lymphoid sheath in the spleen. In the lymph node the

T-cell area is located between the medullary cords with mostly macrophages and plasma cells and lymphoid follicles with mostly B cells [42]. Figure 1.4 shows the spatial distribution of lymphoid cells in the spleen. These cells are located within a few micrometers from each other showing that intercellular distances are very small. It has been shown that suppression by Treg cells is reduced if the number of antigen presenting cells is increased. This finding suggests that a close interaction between Th cells and Treg cells at best at the same antigen presenting cells are required for the suppressive function of Treg cells [81]. Approximately 1000 of 10^5 antigens presented by one antigen presenting cells are of the same specificity [1]. Therefore, interaction at the same antigen presenting cell is possible and interaction between T lymphocytes can assumed to be short ranged. This holds true for paracrine activation of Th cells during an immune response as well as for Th- and Treg-cell interaction. The dependence of the interaction between Th and Treg cells on the distance is analyzed in this work.

1.5 Objectives of this study

The deregulation of the interaction of Th and Treg cells results in development of serious autoimmune diseases. Thus, a better understanding of this interaction is of great importance to control these diseases. The interaction via IL-2 is the foundation of their communication, since the suppressive capacity of Treg cells depends on IL-2 signaling. The interplay between Th cells and Treg cells mediated by IL-2 is defined by the following characteristics (Figure 1.5):

1. In Th cells IL-2 expression is regulated in a network including IL-2 and its receptor. The expression of the IL-2 receptor after antigen stimulation enables autocrine re-uptake of IL-2.

2. Regulatory T cells depend on paracrine IL-2 for their suppressive capacity; Th cells are the main source of IL-2.

3. IL-2 receptor expression includes positive feedback regulation on Th cells and Treg cells. High IL-2 receptor expression on Th cells is associated with Th cell activation and proliferation.

4. Constitutive CD25 expression enables Treg cells to constantly take up IL-2.

5. Regulatory T cells suppress autoimmune reaction by preventing Th cell activation. All differences in IL-2 and the IL-2 receptor expression are mediated by the absence or presence of the transcription factor FOXP3 in Th and Treg cells, respectively.

The molecular basis of IL-2 signaling in Th and Treg cells as well as its consequences on the systemic level are subject of intense investigation. However, the link between these two levels in terms of a mechanistic understanding remains to be determined. Competition for IL-2 between Th cells and Treg cells has been proposed as a mechanism to regulate Th or Treg cell activation. The finding that Treg cell function in cell culture can be mimicked by the addition of antibodies specific for IL-2 supports this hypothesis [17]. However, several open questions remain concerning this competition hypothesis:

1. What are the underlying mechanisms that enable competition for IL-2?

2. Which parameters determine the outcome of competition?

3. Since the hypothesis of competition for IL-2 by Treg cells was formulated, its importance among the described suppressive mechanisms of Th cell activation has been intensively discussed. This analysis may yield a better understanding about the abilities but also limitations of competition for IL-2.

Mathematical modeling is combined with *in vitro* experiments to address these questions. Mathematical modeling enables a detailed analysis of the complex dynamics of the IL-2/IL-2 receptor gene-expression network and allows to study the influence of the intercellular distance on the interaction between Th and Treg cells. The predictions are verified experimentally.

A reaction-diffusion model is established in Chapter 2 that describes the interaction of a Th cell and a Treg cell. It accounts for the IL-2 receptor dynamics of the Th cell as well as the Treg cell and IL-2 secretion, diffusion in the extracellular space and extracellular degradation. Due to the short intercellular distances observed *in vivo* a one-dimensional approach including only the intercellular distance may be serve as a first approximation (cf. Figure 1.4). Variants of this model are used for additional analysis of the IL-2/IL-2 receptor network: the interaction of T helper cells and the interaction of Th cells and Treg cells in the ratio of 2:1.

In Chapter 3 the properties of the IL-2/IL-2 receptor gene-expression network in Th cells are studied in detail. Focus is layed on the role of the autocrine positive feedback on the dynamics as well as the steady-state behavior of the

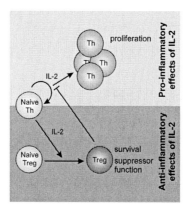

Figure 1.5. The interaction of Th and Treg cells mediated by IL-2 Th cells depend on IL-2 for Th cell activation and a proper ongoing of the immune response. They are also the main source of IL-2 for Treg cells, which suppress Th cell activation.

network. The results are verified in cell culture experiments. In addition, the interaction of Th cells are discussed by means of mathematical model analysis. The differences in the regulation of the IL-2 receptor dynamics of Treg cells are then easily depicted in Chapter 4. Control parameters of this interaction are identified. The mechanisms of Th and Treg cell interaction are verified experimentally. In the outlook first results are presented that describe the effect of Th and Treg cell interaction on the proliferation of Th cells.

Chapter 2

Mathematical Model of T lymphocyte interaction mediated by interleukin 2

The analysis of the IL-2/IL-2 receptor gene-expression network mainly focuses on its structure, but little is known about its dynamical properties. Autocrine positive feedback regulation of the IL-2 receptor as well as the IL-2 expression, activation induced receptor internalization and recycling of the receptor make predictions on the regulation of IL-2 perception very difficult. In this chapter the mathematical model is presented that describes the interaction of a T helper cell and a regulatory T cell to highlight the similarities and differences in the IL-2/IL-2R (IL-2 receptor) gene-expression network of Th and Treg cells. Afterwards variants of this model are introduced.

2.1 Model equations

The IL-2/IL-2R gene-expression network is activated if naive Th cells or resting Th cells are stimulated with their cognate antigen. IL-2 is secreted to the extracellular medium and the expression of IL-2R α-chain enables the Th cell to form the high affinity IL-2 receptor on its surface. The IL-2 receptor dynamics considered in the mathematical model is the dynamics of the α-chain of the IL-2 receptor that is essential to bind IL-2 with high-affinity. On Th cells the IL-2R α-chain is only expressed after antigenic stimulation. Therefore, for the establishment of the IL-2/IL-2 receptor gene-expression network, the number of IL-2R α-chains expressed is assumed to be equal to the number of high-affinity IL-2 receptors. Indeed, for *ex vivo* isolated Th cells it was shown that the number of IL-2R α-chains expressed and the amount of

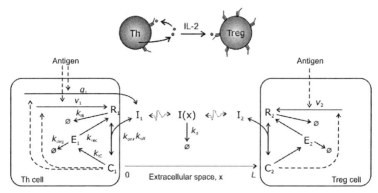

Figure 2.1. Scheme of T helper and regulatory T lymphocyte interaction The reaction-diffusion model accounts for the expression of IL-2 (I, I_n at the plasma membrane) with rate q, its diffusion (D) and extracellular degradation (k_d). The high-affinity IL-2R (R_n) is expressed with the rate v_n. Binding of IL-2 to the IL-2R (k_{on}) leads to the formation of occupied IL-2 receptor (C_n). Its dissociation is characterized by k_{off}. The internalization of R_n and C_n occurs with rate constants k_{iR} and k_{iC}, respectively. R_n is degraded, whereas the internalized IL-2/IL-2R complex (E_n) is either recycled or degraded with rate constants k_{rec} and k_{deg} , respectively. Positive feedback signaling from the occupied IL-2 receptor (C_n) up-regulates the expression of the IL-2 receptor and, in the Th cell, the IL-2 secretion (dashed lines, n =1 for the Th cell and $n = 2$ for the regulatory T cell).

IL-2 bound are strongly correlated [19]. Focusing on the dynamics of IL-2 and the IL-2 receptor expression on the cellular level, the model variables are the IL-2 receptor at the plasma membrane in its unoccupied form (R), as a complex with IL-2 (C) and the endosomal IL-2 receptor after receptor internalization (E). The extracellular IL-2 concentration (I) is described by a reaction-diffusion equation. A one dimensional symmetry is chosen to describe the cellular interfaces. The intercellular distances in cell culture as well as in the lymph nodes are approximately a few micrometers [14], resulting in a dense organization of T lymphocytes. The emerging interfaces are small. Hence, an approximation by a one dimensional symmetry may serve as a first approach. For simplicity and due to limitation in the availability of more detailed experimental data, all processes are modeled with mass-action kinetics except the positive feedback that is modeled for the same reasons as a Hill function. In Figure 2.1, a scheme of the model of an interacting Th cell and Treg cell is presented to illustrate the following equations. The IL-2 receptor dynamics is described by

$$\frac{dR_n}{dt} = v_n - k_{iR}R_n - k_{on}R_nI + k_{off}C_n + k_{rec}E_n \qquad (2.1)$$

$$\frac{dC_n}{dt} = k_{on}R_nI - k_{off}C_n - k_{iC}C_n \qquad (2.2)$$

$$\frac{dE_n}{dt} = k_{iC}C_n - (k_{deg} + k_{rec})E_n \qquad (2.3)$$

where the subscript $n = 1$ represents the Th cell and $n = 2$ the Treg cell. R_n and C_n are the number of unoccupied and occupied IL-2 receptors, respectively. As discussed in section 1.2, the regulation of the IL-2 receptor dynamics in Th cells and Treg cells is identical. The IL-2 receptor expression rate v_n includes a constant rate v_{n0} and a positive feedback dependent term v_{n1}. The positive feedback is modeled as a Hill function with the half saturation constant K and the Hill coefficient m. This assumption is based on the previous reported cooperative binding of STAT5 to the IL-2R α-chain promoter [44].

$$v_n = v_{n0} + v_{n1}\frac{C_n^m}{C_n^m + K^m} \qquad (2.4)$$

Because of the differences in the IL-2 receptor expression of Treg cells mediated by the transcription factor FOXP3, namely their constitutive, high expression of the IL-2 receptor, we assumed that $v_{20} \gg v_{10}$. Furthermore, regulatory T cells are able to enlarge their advantage in IL-2 receptor numbers on their cell surface after stimulation, therefore we set $v_{21} > v_{11}$. All IL-2 receptor expression rates of the T helper cell as well as the regulatory T cell were chosen to match the numbers estimated from flow cytometry. The binding kinetics are described by the association rate constant k_{on} and the dissociation rate constant k_{off}. Due to activation-induced endocytosis, the high affinity IL-2 receptor in its unbound form is internalized with a lower rate constant k_{iR} than in a complex with IL-2 (k_{iC}). Unlike the unoccupied receptor (R_n) the internalized IL-2/IL-2 receptor complex (E_n) can be recycled to the membrane with the rate constant k_{rec}. Endosomal degradation of the IL-2/IL-2 receptor complex is assumed to occur with the rate constant k_{deg}. The values of all model parameters are given in Table 2.1 at the end of this chapter.

The IL-2 dynamics is described by a reaction-diffusion equation with the diffusion coefficient D and the extracellular degradation rate constant k_d

$$\frac{\partial I}{\partial t} = -k_d I + D\frac{\partial^2 I}{\partial x^2}. \tag{2.5}$$

The IL-2 dynamics at the plasma membrane of the Th cell is given by mixed boundary conditions (Eq. 2.6) describing IL-2 secretion (q), binding (k_{on}) and dissociation (k_{off}) of its receptor. It is assumed that the plasma membrane of the Th cell is located at $x = 0$ and the plasma membrane of the Treg cells at $x = L$. Regulatory T cells are unable to produce IL-2 (Eq. 2.7), therefore only the receptor interaction of IL-2 is considered:

$$-D\frac{\partial I}{\partial x}\bigg|_{x=0} = q - k_{on}R_1 I_{x=0} + k_{off}C_1 \tag{2.6}$$

$$-D\frac{\partial I}{\partial x}\bigg|_{x=L} = -k_{on}R_2 I_{x=L} - k_{off}C_2 \tag{2.7}$$

IL-2 expression is induced by antigenic stimulation (q_0) and augmented by autocrine signaling through the IL-2 receptor (q_1). This has been inferred from the reduction in IL-2 expressing cells, when Th cells are stimulated in the presence of anti-IL-2 antibodies (cf. Figure 1.1A, [17]). Therefore, the IL-2 secretion rate (q) includes a constant secretion rate q_0 that depends on the strength of the T cell receptor stimulus, and a feedback dependent rate q_1. As explained above, cooperativity is also assumed for the IL-2 receptor-induced IL-2 gene-expression

$$q = q_0 + q_1 \frac{C_1^m}{C_1^m + K_q^m}, \qquad (2.8)$$

where K_q is the half saturation constant of the Hill function and m the Hill coefficient. In summary, the model equations describing the IL-2/IL-2 receptor gene-expression network of Th cells and regulatory T cells are very similar. The difference in IL-2 secretion, namely that Treg cells are unable to produce IL-2, is described in the different boundary conditions for the two cells types (cf. Equation 2.6 and 2.7). The initially high number of IL-2 receptors on Treg cells independent of antigenic stimulation is accounted for by the much higher value of the IL-2 receptor expression rate v_{20} (Treg cell) compared to v_{10} (Th cells, cf. Table 2.1).

2.2   Quasi-steady-state approximation due to fast diffusion

In general, diffusion is a very fast process taking place in the order of seconds, whereas changes in the IL-2 receptor dynamics are only observed within hours (cf. Figure 3.1). The characteristic time with which IL-2 equilibrates in the extracellular medium is approximated using a simplified model describing the extracellular IL-2 dynamics (Appendix A). The leading spatially inhomogeneous mode of the solution for the spatio-temporal IL-2 concentration relaxates in less than 1 s, e.g., for a distance of 10 µm and an re-uptake by 5000 receptors per cell. By contrast, the characteristic time for IL-2R α-chain expression is 6-10 h. In addition to this clear separation of time scales of IL-2 diffusion and the IL-2 receptor expression, IL-2 binds to its receptor with high affinity (10^{-11} M) [42]. We can therefore assume IL-2 diffusion to be in a steady state on the time scale of the IL-2 receptor dynamics:

$$\frac{\partial I}{\partial t} = 0 \qquad (2.9)$$

Comparing this approximation with direct numerical solution of the full system, the approximation is found to be reasonably accurate for the distance L up to several hundred µm [83]. Using Equations 2.5-2.7, the IL-2 concentrations at the cell surfaces of the Th cell is obtained as

$$I_1(t) \;=\; \frac{(\psi + k_{on}R_2\chi)(q + k_{off}C_1) + 2k_{off}C_2\psi\phi}{k_{on}^2 R_1 R_2 \chi + k_{on}\psi(R_1 + R_2) + \chi\psi^2} \tag{2.10}$$

$$\chi \;=\; \tanh\!\left(L\sqrt{\frac{k_d}{D}}\right) \tag{2.11}$$

$$\phi \;=\; \frac{e^{\sqrt{\frac{k_d}{D}}L}}{e^{2\sqrt{\frac{k_d}{D}}L} + 1} \tag{2.12}$$

$$\psi \;=\; N_A F \sqrt{Dk_d} \tag{2.13}$$

and for the IL-2 concentration at the plasma membrane of the regulatory T cell as

$$I_2(t) = \frac{2\psi(q + k_{off}C_1)\phi + k_{on}k_{off}R_1 C_2 \chi + k_{off}C_2\psi}{k_{on}^2 R_1 R_2 \chi + k_{on}\psi(R_1 + R_2) + \chi\psi^2} \tag{2.14}$$

To be able to specify the IL-2 secretion and the IL-2 receptors in the unit number per cell, the Avogadro constant (N_A) and the cell surface (F) appear explicitly in the equations. Inserting $I_1(t)$ and $I_2(t)$ into the equation describing the IL-2 receptor dynamics (Eq. 2.1-2.3), the problem is reduced to a system of ordinary differential equations, which enables us to perform extensive steady-state analysis.

2.3   Model variants

The model presented in the previous sections describes the two interacting IL-2/IL-2 receptor gene-expression networks of a Th cell and a Treg cell. Variants of this model are analyzed. Figure 2.2 gives an overview over all models considered. The equations for the IL-2 receptor dynamics and the reaction-diffusion equation are identical for all model variants. The differences are described in the boundary conditions. To study the dynamics of the IL-2/IL-2 receptor gene-expression network and the extracellular IL-2 distribution in detail and separately from the intercellular interaction, a model is analyzed describing this network in a single Th cell. It is assumed that the diffusion of IL-2 takes place in a finite extracellular space. A zero flux boundary condition is considered at the diffusion barrier (Figure 2.2).

A model describing the interaction of two Th cells is used to study the autocrine and paracrine action of IL-2. Both cells are able to secrete IL-2. As in

Equation 2.8 the IL-2 secretion rate (q_n) includes an antigen dependent (q_{n0}) and a feedback dependent term (q_{n1}), which are not necessarily equal. For both model variants the quasi-steady-state approximation is applied. The IL-2 concentrations at the plasma membrane are given in the Appendix B. A model of two Th cells and one regulatory T cell is also considered to analyze the interaction of Th cells and Treg cells in a situation different from the equal distribution. But since it is a combination of the models presented here, it will be discussed shortly in Chapter 4 and in the Appendix B.3.

2.4 Model parameters

The majority of the model parameters has been taken directly or estimated from literature. From flow cytometry analysis we estimated the number of cell-surface IL-2R α-chain molecules to be of the order of 1000-2000 per unstimulated regulatory T cell [17]. Following stimulation, IL-2R α-chain expression on Treg cells can increase by about one order of magnitude. Under conditions of maximal stimulation, Th cells also express high amounts of IL-2R α-chain. However, their expression remain by a factor of 3 below the peak values seen on Treg cells. Knowing the internalization rate constants these estimates can be used to determine the expression rates of the IL-2 receptor. The internalization of the IL-2 receptors has been studied in the human T cell line, which constitutively expresses the IL-2 receptor. The internalization rate constants the of unoccupied IL-2 receptor and the IL-2/IL-2 receptor complex $(k_{iR}$ and $k_{iC})$ were approximately 0.64 h^{-1} and 1.7 h^{-1}, respectively [22]. Given these internalization rates, the expression rates of IL-2R α-chain under basal and antigen-stimulated conditions on Th cells (v_{10}, v_{11}) and Treg (v_{20}, v_{21}) were chosen so as to match the estimated receptor numbers per cell (Table 2.1). Kum et al. [46] measured IL-2 secreted by human peripheral blood cells. From their data, we estimated IL-2 secretion rates of approximately 23400 molecules/cell/h in the first 8 h of antigen stimulation and 6000 molecules/cell/h between 8 h and 12 h, assuming that every cell produces IL-2. In the model, we varied the IL-2 secretion rate due to antigenic stimulation (q_0) between 0 and 22000 molecules/cell/h. Although IL-2 expression is primarily induced by antigenic stimulation (q_0), it is also augmented by autocrine signaling through the IL-2 receptor (q_1). This has been inferred from the reduction of IL-2 expressing cells, when Th cells are stimulated in the presence of anti-IL-2 antibodies [17]. To account for the relative importance of the antigen induced and feedback dependent IL-2 secretion rate, we assumed a rather small value for q_1 and set $q_1 = 1000$ molecules/cell/h.

IL-2 receptor dynamics

$$\frac{dR_n}{dt} = v_n - k_{iR}\,R_n - k_{on}\,R_n\,I_n + k_{off}\,C_n + k_{rec}\,E_n \qquad v_n = v_{n0} + v_{n1}\,\frac{C_n{}^m}{K^m + C_n{}^m}$$

$$\frac{dC_n}{dt} = -k_{iC}\,C_n + k_{on}\,R_n\,I_n - k_{off}\,C_n$$

$$\frac{dE_n}{dt} = k_{iC}\,C_n - (k_{rec} + k_{deg})\,E_n$$

IL-2 dynamics

$$\frac{\partial I}{\partial t} = -k_d I + D\,\frac{\partial^2 I}{\partial x^2} \approx 0$$

Th and Treg cell interaction

$$-D\,\frac{\partial I}{\partial x}\bigg|_{x=0} = q - k_{on}R_1 I + k_{off}C_1$$

$$-D\,\frac{\partial I}{\partial x}\bigg|_{x=L} = k_{on}R_2 I - k_{off}C_2$$

$$v_2 > v_1$$

Single Th cell

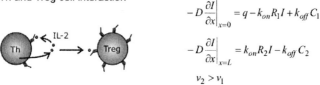

$$-D\,\frac{\partial I}{\partial x}\bigg|_{x=0} = q - k_{on}R_1 I + k_{off}C_1$$

$$-D\,\frac{\partial I}{\partial x}\bigg|_{x=L} = 0$$

Th/Th cell interaction

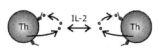

$$-D\,\frac{\partial I}{\partial x}\bigg|_{x=0} = q_1 - k_{on}R_1 I + k_{off}C_1$$

$$-D\,\frac{\partial I}{\partial x}\bigg|_{x=L} = -q_2 + k_{on}R_2 I - k_{off}C_2$$

$$q_n = q_{n0} + q_{n1}\,\frac{C_n{}^m}{K_q{}^m + C_n{}^m}$$

Figure 2.2. Overview of model equations and model variants The dynamics of the IL-2 receptor is common for all model variants. The differences of the models describing the Th and Treg cell interaction, the single Th cell model and the model of two interacting Th cells are given in an overview. The properties of the Th/Treg cell model are discussed in Chapter 2.1. For the model of a single Th cell the extracellular space is limited by a diffusion barrier (zero flux boundary condition at $x = L$). Analyzing the interaction of two Th cells, a IL-2 secretion rate is considered for both cells. ($n = 1$ for the Th cell and $n = 2$ for the regulatory T cell or the second Th cell)

Secreted IL-2 binds to its high-affinity receptor with the association and dissociation rate constant (k_{on} and k_{off}) of 111.6 nM^{-1}h^{-1} and 0.83 h^{-1}, respectively [86]. The diffusion coefficient (D) of IL-2 in cell culture (and *in vivo*, e.g., in lymph nodes) is not known precisely, but is certainly less than its value in water (349 x 10^3 µm^{2}h^{-1}[26]). We have chosen 36000 µm^{2}h^{-1}. We also included degradation of the extracellular IL-2 in our model to account for the loss of IL-2 in the extracellular medium. However, the model indicates that IL-2 removal is predominantly due to cellular uptake and endosomal degradation; the extracellular degradation constant k_d is set to 0.1 h^{-1}. In the case of no IL-2 withdrawal by other cells we set k_d to 10 h^{-1} for the analysis of a single Th cell. The internalized IL-2/IL-2 receptor complex is sorted in the endosomal pool; IL-2R α-chain are recycled to the cell surface [36] with the rate constant (k_{rec}) or degraded (k_{deg}). We set k_{rec} to 9 h^{-1}, which is approximately double the time measured for Transferrin in Hep2 cells resulting in an overall half life of approximately 9.5 min in the sorting and recycling endosomes [32]. Contradictory results about the endosomal degradation of IL-2R α-chain have been reported. Duprez and Dautry-Varsat [23] observed no reappearance of the receptor at the plasma membrane, whereas measurements of Hemar et al. [36] correspond to a recycling probability for IL-2R α-chain of 90%. We chose an intermediate recycling IL-2R α-chain probability and set the endosomal degradation rate constant (k_{deg}) to be 5 h^{-1}. In contrast to IL-2R α-chain, IL-2 is not recycled, but degraded [23, 24, 36].

Since little quantitative information is available on IL-2 dependent enhancement of the IL-2R α-chain expression and IL-2 secretion a simple Hill-type response function with a half-saturation constant $K_1 = K_2 = K_q = 1000$ occupied receptors per cell and a Hill coefficient $m = 3$ was chosen. Sigmoid response functions may arise through a variety of molecular mechanisms, e.g. cooperativity [69]. For STAT5, which mediates IL-2 signaling, cooperativity is reported for its binding to the IL-2R α-chain promoter [44]. In this work, the number ($\#$) per cell and the concentration nM are chosen as standard units for the IL-2 receptor and the IL-2 concentration, respectively. The area of the cell surface is assumed to be 300 μm^2. The parameter values are summarized in Table 2.1 and taken as the standard set of parameters throughout the analysis of this work. The antigen induced IL-2 secretion rate (q_0) is used as a control parameter representing the strength of the antigenic stimulation and is varied as indicated. Likewise, the intercellular distance is varied when its influence on the cellular interaction is studied.

Parameter	Symbol	Value
IL-2 dynamics		
Antigen stimulated IL-2 secretion rate	q_0	0-22000 molec.cell^{-1}h^{-1}
Feedback induced IL-2 secretion rate	q_1	1000 molec. cell^{-1}h^{-1}
Half saturation constant of the feedback on the IL-2 expression	K_q	1000 molec. cell^{-1}
Hill coefficient	m	3
Diffusion coefficient of IL-2	D	36000 μm^{2}h^{-1}
Rate constant of extracellular IL-2 degradation	k_d	0.1 h^{-1} or 10 h^{-1}
IL-2 receptor dynamics		
Antigen stimulated IL-2 receptor expression rate (Th)	v_{10}	150 molec. cell^{-1}h^{-1}
Feedback induced IL-2 receptor expression rate (Th)	v_{11}	3000 molec. cell^{-1}h^{-1}
Antigen stimulated IL-2 receptor expression rate (Treg)	v_{20}	1000 molec. cell^{-1}h^{-1}
Feedback induced IL-2 receptor expression rate (Treg)	v_{21}	8000 molec. cell^{-1}h^{-1}
Half-saturation constant of feedback expression of IL-2Rα	$K_{1,2}$	1000 molec. cell^{-1}
IL-2 association rate constant to IL-2R	k_{on}	111.6 nM^{-1}h^{-1}
IL-2 dissociation rate constant from IL-2R	k_{off}	0.83 h^{-1}
Internalisation rate constant of IL-2R	k_{iR}	0.64 h^{-1}
Internalisation constant of IL-2/IL-2R complex	k_{iC}	1.7 h^{-1}
Recycling rate constant of IL-2Rα	k_{rec}	9 h^{-1}
Endosomal degradation constant IL-2R	k_{deg}	5 h^{-1}
Spatial parameters		
Cell surface area	F	300 μm^2
Cell-to-cell distance	L	10-1000 μm

Table 2.1: **Model parameters**

Chapter 3

The IL-2/IL-2 receptor gene-expression network

In this chapter, the IL-2/IL-2 receptor network is introduced. Of special interest is its response to different strength of antigenic stimulation as a control parameter for cellular communication. The expression of IL-2 and the IL-2R α-chain gene is among the first events after the stimulation of Th cells with their cognate antigen. The expression of these two genes is regulated in a network that involves an autocrine positive feedback loop of the IL-2R α-chain and the IL-2 gene-expression, activation-induced internalization and recycling of the receptor. Autocrine IL-2 signaling is required for both Th cell proliferation and differentiation. Focusing on a single Th cell the IL-2/IL-2 receptor expression network is studied here in detail.

3.1 The dynamics of the IL-2/IL-2 receptor gene-expression network

Th cells are exposed to different strength of antigenic stimulation. Already on the level of TCR signaling several mechanisms are described that can distinguish between weak and strong signals [1, 16]. On the level of gene-expression cooperative phosphorylation and dimerization of the transcription factor NFAT controls IL-2 secretion. A correlation between the strength of the antigen stimulus and the amount of IL-2 secreting cells is reported [69, 60, 37]. The strength of antigenic stimulation also corresponds to the percentage of cells that express high amounts of the IL-2R α-chain, as the kinetics of IL-2R α-chain expression in *ex vivo* isolated Th cells for a moderate

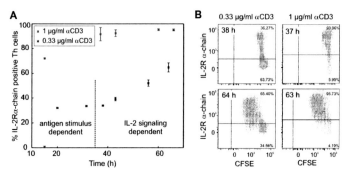

Figure 3.1. **Upregulation of the IL-2R α-chain** (A) CD4$^+$CD25$^-$ Th cells were isolated from C57/Bl6 mice and stimulated with 0.33 µg/ml (blue) or 1µg/ml (red) anti-CD3 and APCs. For a strong antigen stimulus (1µg/ml) approximately all cells are activated at the time of maximal IL-2 secretion (approx. 20h, cf. Figure 1.1). At 0.33 µg/ml, only a fraction of the Th cells is activated, which increases after the induction of proliferation. (B) Proliferating cells are shown at approximately 38h and 64 h after stimulation (cf. A). At 38 h cells stimulated with 0.33 µg/ml anti-CD3 only start to proliferate, whereas cells stimulated with 1µg/ml anti-CD3 passed already several round of proliferation.

(0.330 µg/ml anti-CD3) and strong (1µg/ml anti-CD3) antigen stimulation reveals (Figure 3.1A). Strong antigen stimulus results in the activation of almost all cells around the time of maximal IL-2 secretion (20h, cf. section 1.1) and expression stays high over several days. Interestingly, for the stimulation with 1 µg/ml anti-CD3 two situations are observed at 15 h after stimulation. In one situation the expression level of the IL-2R α-chain is high in the majority of the cells, whereas in the other situation almost no cells are activated. The strong upregulation of the IL-2 receptor is followed by rapid proliferation shown in Figure 3.1B. Proliferation is measured by CFSE (carboxyfluorescein succinimidyl ester) staining. CFSE binds intracellularly to the plasma membrane and its concentration per cell is approximately halved every time the cell divides. This results in multiple population with distinct CFSE concentration, each associated to one generation. The moderate stimulus (0.330 µg/ml anti-CD3) results only in the activation of a fraction of cells and little proliferation, whereas with the strong stimulus the majority has started to proliferate at approximately 38 h (Figure 3.1A and B). Hence, the number of activated Th cells as well as the time to first division are controlled by the strength of the antigen stimulus.

The mathematical model describing the IL-2/IL-2 receptor gene-expression

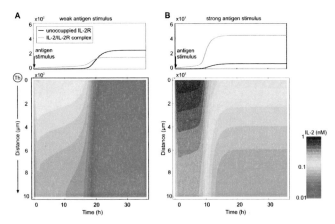

Figure 3.2. Autocrine positive feedback regulates extracellular IL-2 The IL-2R kinetics is shown for the unoccupied (black) and the occupied IL-2R (gray) for low IL-2 secretion rate (A, $q_0 = 2000$ molecules/h) and strong IL-2 secretion rate (B, $q_0 = 8000$ molecules/h). The upregulation of the IL-2R is accompanied with a rapid drop in extracellular IL-2 for both IL-2 secretion rates (bottom panel). High IL-2 concentration is localized to the plasma membrane ($k_d = 0.1/h$).

network of a single cell is used to address the question how the strength of the antigen stimulus is translated into the amount of IL-2R α-chain expressing cells. As stated in section 1.1, the IL-2R α-chain expression mainly depends on its own expression, although crosstalk between the T cell receptor signaling and the IL-2 receptor signaling is likely to occur [40].

In Figure 3.2A and B, the kinetics of the occupied and unoccupied IL-2 receptor and the extracellular distribution of the IL-2 concentration are shown for a small and a high antigen-induced IL-2 secretion rate. The upregulation of the IL-2 receptor occurs rapidly after a period of time with low IL-2 receptor expression. The time delay until the IL-2 receptor is upregulated as well as the amount of IL-2/IL-2 receptor complexes depend on the strength of the antigen stimulus (cf. Figure 3.2A and B). The spatio-temporal distribution of the extracellular IL-2 concentration is shown below its corresponding IL-2 receptor kinetics. The upregulation of the IL-2 receptor is accompanied by a rapid drop in extracellular IL-2 concentration. This finding suggests that the IL-2 receptor expression depends on the strength of the antigen stimulus via the amount of secreted IL-2.

Almost no extracellular IL-2 is detected in the extracellular medium after

IL-2 receptor upregulation for the small IL-2 secretion rate (Figure 3.2A). In the case of high IL-2 secretion the IL-2 concentration is also strongly reduced as soon as the IL-2 receptor expression is high due to autocrine uptake of IL-2 (Figure 3.2B). The positive feedback regulation and also the recycling of the IL-2 receptor provide an activation-induced supply of unoccupied receptors. Thus, high IL-2 concentration in the order of nM occurs only localized at the plasma membrane and practically no IL-2 is detected at a distance of 100 µm. This finding is in agreement with experimental measurements of very low IL-2 concentration in the supernatant of cultured Th cells. These measurements revealed a very low IL-2 concentration of a few pM [61], which was unexpected since the IL-2 secretion rate has been determined to be high [46]. Our simulation suggests that the high uptake rate of its receptor localizes the IL-2 concentration efficiently resulting in these measured low concentration in the bulk phase.

In summary, the activation of the autocrine positive feedback loop is required for high IL-2 receptor expression and therefore for Th cell activation. IL-2 accumulates in the extracellular medium until a threshold concentration is reached to activate the feedback loop. This finding suggests that the observed dependence of the amount of activated Th cells and also the time until the first division on the strength of the antigen stimulation is mediated by its regulation of the extracellular IL-2 concentrations. Although it is reported that the initiation of proliferation is mainly antigen dependent [43], this analysis rather suggests that the activation of the autocrine positive feedback loop and subsequent the strength of IL-2 signaling is crucial for its initiation, at least in the case of moderate antigen stimulation (Figure 3.2A). In addition to the dependence of Th cell activation on autocrine re-uptake of IL-2, it localizes the extracellular IL-2 concentration to the proximity of the secreting cell. Autocrine positive feedback regulation, activation induced receptor endocytosis and receptor recycling ensure this strong localization. In the following the focus will be on the steady-state analysis for further understanding of the regulation of Th cell activation.

3.2   Positive feedback regulation of the IL-2 receptor causes bistability

In the previous section the experimental analysis revealed a dependence of the amount of IL-2R α-chain expressing cells on the strength of the antigen stimulus. In this section we address the underlying mechanism. In Figure

3.3A, the steady-state concentration of the IL-2 receptor is shown in dependence on the antigen-induced IL-2 secretion rate (q_0). The dose- response
curve is bistable. The IL-2 receptor expression rate is low as long as the IL-2
secretion rate is below the activation threshold, θ. At the activation threshold, a saddle-node bifurcation occurs. After the activation threshold has
been passed, the IL-2 receptor expression switches to a high state. A further
increase in the strength of the antigen-induced IL-2 secretion rate does not
lead to an increase in the IL-2 receptor numbers. After saturation of the autocrine positive feedback-induced expression, the IL-2 receptor numbers are
maintained by a balance of IL-2 receptor recycling and activation-induced
receptor endocytosis (Figure 3.3A). As expected, an increase in the IL-2 secretion rate leads to an accumulation of extracellular IL-2. However, the
extracellular IL-2 concentration in the steady state stays low over a wide
range of IL-2 secretion rates confirming the efficient autocrine uptake of IL-2
(Figure 3.3B).

The expression of the IL-2 receptor and the extracellular IL-2 concentration
exhibit bistability. Bistability is characterized by a parameter region in which
two stable steady states coexist. Their area of attraction is separated by an
unstable steady state. Interestingly, the difference in the thresholds of the
IL-2 secretion rates for up and down-regulation of the IL-2 receptor, also
known as hysteresis, is not very pronounced. This reveals a strong coupling
between IL-2 secretion and IL-2 receptor expression (cf. also Appendix C).

The state of low IL-2 receptor expression can be interpreted as the naive
state of Th cells, which is characterized by the absence of high-affinity IL-
2 receptors. In contrast, activated Th cells express a high number of IL-
2 receptors on their surface (Figure 3.3A). The autocrine positive feedback
loop, thus controlling the IL-2 receptor expression, causes a translation of
the continuous antigen stimulus into a digital all-or-nothing response of Th
cell activation.

The bistable IL-2 receptor expression pattern of a single cell should result in a
bimodal distribution of cells with low and high IL-2 receptor expression in cell
culture, if the stimulus strength is close to the activation threshold. Various
sources of variability exist that could cause this distribution among which the
variability in the strength of the antigen stimulation and the heterogeneity
in the extracellular IL-2 concentration can be assumed to predominate. This
variability normally results in a log-normal distribution of protein numbers
[50], as also observed for the number of IL-2 receptor on the cell surface
[76]. We assume, therefore, that the expression rate of the IL-2 receptor
is log-normally distributed around the parameter value given in Table 2.1.
This assumption indeed results in a bimodal expression pattern of the IL-2

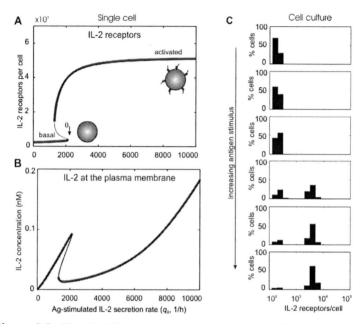

Figure 3.3. Bistable IL-2 receptor expression due to its autocrine positive feedback regulation (A) The stimulus-response curve of the IL-2 receptor exhibits a bistable expression pattern. (The stable steady states are marked with thick the unstable states with thin lines.) The IL-2 receptor expression remains in a basal, IL-2 receptor low state up to a critical IL-2 secretion rate (θ). For stimuli beyond this threshold, the IL-2 receptors attain an activated level, which is sustained by positive feedback of the IL-2 receptor expression. (B) The stimulus-response curve of IL-2 concentration at the plasma membrane reveals that its concentration remains low for various antigen induced IL-2 secretion rate. (C) Simulated IL-2 receptor expression in a Th cell population. In an ensemble of 1000 cells the rate constant of the antigen- and feedback-dependent IL-2 receptor expression were log-normally distributed ($v_{00} = 150 \pm 23$ mol./cell/h and $v_{10} = 3000 \pm 762$ molecules/cell/h). The distributions were computed for increasing levels of antigen-induced IL-2 secretion rates ($q_0 = 0$, 300, 600, 1800, 2700, 15000 molecules/cell/h). The discrete switching in a single cell translate into a binary response pattern of the cell population.

receptor in an ensemble of 1000 cells. With increasing antigen-induced IL-2 secretion rates the percentage of cells expressing a high amount of IL-2 receptors increases (Figure 3.3C). Simulating Th cells with different antigen concentrations should give comparable distributions in cell culture. Before testing this prediction in cell culture experiments, the mathematical model is used to analyze the sensitivity of the bistable behavior to changes in the parameter values.

3.3   Analysis of a simplified model

To understand how bistability emerges through autocrine IL-2 signaling, we considered a simplified model of one Th cell secreting IL-2 into a finite extracellular volume. It is further assumed that internalized IL-2/IL-2 receptor complexes are completely degraded. The differential equations read as followed:

$$\frac{dR}{dt} \;=\; v - k_{iR}R - k_{on}RI + k_{off}C \tag{3.1}$$

$$\frac{dC}{dt} \;=\; k_{on}RI - k_{off}C - k_{iC}C \tag{3.2}$$

$$\frac{\partial I}{\partial t} \;=\; -k_d I + D\frac{\partial^2 I}{\partial x^2} \tag{3.3}$$

where variables and parameters have the same meaning as in the full model. The boundary conditions describe IL-2 secretion and binding to IL-2 receptors

$$-D\frac{\partial I}{\partial x}\bigg|_{x=0} = q - k_{on}R_1 I + k_{off}C_1 \tag{3.4}$$

and an inert boundary at the other end of the volume,

$$-D\frac{\partial I}{\partial x}\bigg|_{x=L} = 0 \tag{3.5}$$

so that IL-2 is either degraded in the extracellular space or through re-uptake by the cell. Applying the quasi-steady-state approximation as described

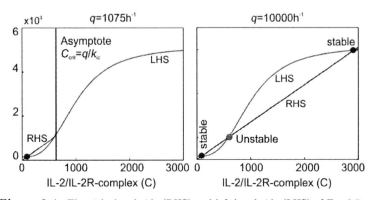

Figure 3.4. The right-hand side (RHS) and left-hand side (LHS) of Eq. 3.7 are plotted as functions of C. The intersection points mark the steady states (stable and unstable as indicated). The number of steady states depends on the value of the IL-2 secretion rate q. The left graph depicts the critical situation when two new steady states emerge through a saddle-node bifurcation.

---

above, the IL-2 concentration at the plasma membrane of the Th cell is obtained from Eqs. 3.3-3.5 as

$$I(0,t) = \frac{q + k_{off}C}{N_A F \sqrt{k_d D} \tanh(L\sqrt{k_d/D}) + k_{on}R} \tag{3.6}$$

Inserting Eq. 3.6 into Eqs. 3.1 and 3.2 and taking into account the dependence $v(C)$, we obtain the following implicit equation for the steady state of the occupied IL-2 receptors, C:

$$v_0 + v_1 \frac{C^m}{C^m + K^m} = \left(\frac{\vartheta}{q - k_{iC}} + k_{iC} \right) C \tag{3.7}$$

where we have introduced the parameter combination:

$$\vartheta = \frac{k_{iR}(k_{iC} + k_{off})}{k_{on}} N_A F \sqrt{k_d D} \tanh(L\sqrt{\frac{k_d}{D}}).$$

Moreover, we find as condition for non-negative values of unoccupied receptors (R)

$$C < \frac{q}{k_{iC}}. \tag{3.8}$$

In Eq. 3.7 the effective production (l.h.s.) and degradation (r.h.s.) terms are plotted as functions of C. Depending on the value of q, there are either one or three intersections of the two curves in the admissible range of C (obeying Eq. 3.8), as shown in Figure 3.4. There can be either one steady state (stable) or three steady states (two stable states, separated by an intermediate unstable state).

As seen from Figure 3.4 for small values of the IL-2 secretion rate q, there is only one steady state with small C, which also implies that there is no autocrine upregulation of IL-R α-chain. As q increases, the effective degradation rate for C decreases, and the range of admissible C values, bounded by Eq. 3.8, increases. As shown in Figure 3.3, at a critical value of q a saddle-node bifurcation takes place, creating a stable steady state of upregulated IL-2R α-chain supported by autocrine feedback. To be able to affect a switch from basal to activated IL-2R α-chain expression by increasing q, the (unstable) intermediate and (stable) low steady state of C must merge in a second saddle-node bifurcation. For sufficiently large values of q, Eq. 3.7 becomes approximately:

$$v_0 + v_1 \frac{C^m}{C^m + K^m} = k_{iC} C \qquad (3.9)$$

Whether this equation has one or three solutions for C depends on the kinetic parameters. In particular, for only a single, activated steady state to exist there must be a non-zero basal rate of IL-2R α-chain expression v_0. In the real system, this part is played by the antigen-activated IL-2R α-chain expression. Moreover, this simplified analysis indicates that the positive-feedback regulation of IL-2R α-chain is sufficient to obtain bistability and positive feedback to IL-2 secretion is not required.

3.4 Parameter dependence of bistability

To analyze the sensitivity of the activation threshold θ the parameters describing the IL-2 receptor dynamics and the IL-2 signaling are varied to half and double the values given in Table 2.1 (Figure 3.5). The changes in the parameter values describing the IL-2 receptor dynamics (left panel) reveal a strong dependence of the activation threshold on these parameters. Equally high control has the half-saturation constant K (Figure 3.5, right panel). This constant gives a measure how many IL-2/IL-2 receptor complexes are required to activate the autocrine positive feedback loop. Thus the properties of the IL-2 receptor dynamics and the positive feedback determine the

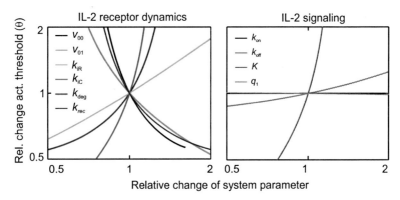

Figure 3.5. Parameter-sensitivity analysis The activation threshold (θ) strongly depends on the IL-2 receptor dynamics and the autocrine positive feedback regulation.

critical IL-2 concentration that is required for Th cell activation. As already described in the previous section the antigen-induced IL-2 receptor secretion rate (v_{00}) has to be greater than zero to obtain bistability. It is also nicely seen that the positive feedback regulation of the IL-2 expression (q_1) has no influence of the activation threshold as reported above.

Since autocrine IL-2 signaling is crucial for Th cell activation, extracellular properties such as extracellular degradation or unspecific binding of IL-2 to neighboring cells and the size of the extracellular space are assumed to influence the re-uptake of IL-2 by the Th cell. The extracellular environment is represented in the model by the parameter describing the extracellular space as a distance from the plasma membrane (x). At the distance L a diffusion barrier is assumed. The extracellular degradation rate constant of IL-2 (k_d) and the diffusion coefficient (D) determine the loss of IL-2 at the plasma membrane. The dependency of the region of bistability (gray region) on the extracellular IL-2 degradation rate constant (k_d) and the size of the extracellular space (L) is given in Figure 3.6. For extensions of the extracellular space below 10-15 µm, bistability exists for a wide range of extracellular degradation rate constants (k_d). Small distances are shown to oc-

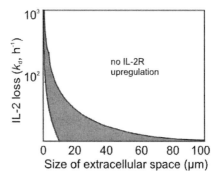

Figure 3.6. **Bistable region in dependence of the spatial parameters**
For a moderate antigen induced IL-2 secretion rate ($q_0 = 3000$ molecules/h) the
region in the parameter space, where bistability exists (gray region), is shown in
dependence of the extracellular IL-2 degradation rate constant (k_d) and the distance
to the plasma membrane (L). For small distances the bistable region exists for a
great range of the IL-2 degradation rate constant.

cur *in vivo* (Figure 1.4) and can also assumed under cell culture conditions.

3.5 Experimental test of all-or-nothing activation of T helper cells

In Figure 3.7A (cf. Figure 3.3C), it has been demonstrated that in cell
culture due to natural occurring variability this bistable behavior results
in a bimodal distribution of the IL-2 receptor expression. With increasing
strength of antigen stimulation the amount of cells with high IL-2 receptor
expression increases.

For experimental validation, CD4$^+$CD25$^-$ T lymphocytes from OVA-TCR$^{tg/tg}$
DO11.10 mice were isolated. (These mice carry a trans-gene coding for a
TCR with known specificity (Ova-peptide). *Ex vivo* isolated T cells from
these mice allow TCR stimulation with antigen in cell culture.) In separate
experiments the cells were stimulated with different concentrations of their
cognate antigen (OVA). After 72 hours of culture cells were analyzed for IL-
2R α-chain expression by FACS-analysis. In Figure 3.7B, the IL-2 receptor
expression profile is shown for the different antigen concentrations used. The
distribution of IL-2 receptor expression is clearly bimodal. In agreement with

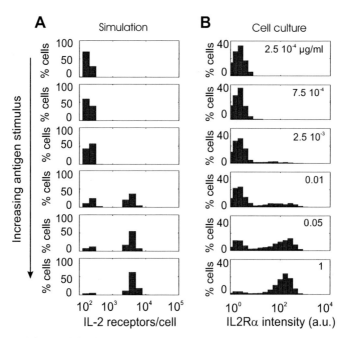

Figure 3.7. Positive feedback causes all-or-nothing activation of Th cells
(A) Bimodal IL-2 receptor expression is observed simulation an ensemble of 100 cells with log-normally distributed IL-2 receptor expression rates (cf. Figure 3.3C). (B) Binary IL-2R α-chain expression pattern in Th-cell populations stimulated with increasing doses of cognate antigen. OVA-TCR[tg/tg]CD4[+]CD25[-]T cells from DO.11.10 mice were cultured with irradiated antigen-presenting cells and subjected to different concentration of OVA peptide (2.5×10^{-4}, 7.5×10^{-4} 2.5×10^{-3}, 0.01, 0.05 and 1 μg/ml), surface expression of the IL-2R α-chain was determined by flow cytometry after 72 hours. The distribution were generated by appropriate binning of the IL-2R α-chain fluorescence intensities.

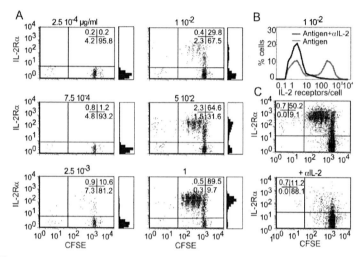

Figure 3.8.   Positive feedback causes all-or-nothing activation of Th cells (A) IL-2R α-chain expression (cf. Figure 3.7) and proliferation were determined for the OVA-peptide concentration indicated in Figure 3.7. Labeling the Th cells with the proliferation marker CFSE reveals that proliferation is strongly correlated with high expression of the IL-2R α-chain. (B) Addition of blocking anti-IL-2 antibodies to the cell culture (black line) completely prevents the IL-2R α-chain up-regulation observed without any antibodies present (gray line). (C) Proliferation is driven by the positive feedback. Addition of blocking antibodies for IL-2 abolished cell proliferation (right figure) and occurs under control condition (left figure). (Data was kindly provided by Alexander Scheffold, DRFZ, Berlin.)

the simulation, the percentage of cells with high IL-2R α-chain expression increases with the concentration of OVA-peptide.

In addition, we analyzed the dependence of the strength of antigen stimulation on Th cell proliferation (CFSE staining). Figure 3.8A reveals that cell proliferation is linked to high IL-2 receptor expression. Thus, only for moderate and high OVA-peptide concentrations (0.01-1 µg/ml OVA-peptide) proliferation is observed.

The model proposes that the bimodal expression pattern of the IL-2 receptor is due to the autocrine positive feedback loop. To test this prediction blocking antibodies specific for IL-2 were added to the culture. This interruption of the autocrine positive feedback should prevent Th cell activation. In Figure 3.8B, it is shown that the addition of the antibodies completely inhibits the expression of the IL-2 receptor (black line), whereas in the control experiment at the same concentration of OVA-Peptide a clear bimodal distribution is observed (gray line). Therefore, the bimodal distribution of IL-2 receptor expression as well as its dependence on the autocrine positive feedback loop could be validated experimentally.

It has been shown previously that the time period between the onset of stimulation and proliferation depends on the strength of antigenic stimulation (Figure 3.1B). Simulations of the IL-2/IL-2 receptor network suggest that this is due to the different times needed to activate the autorine positive feedback loop. In Figure 3.8C the proliferation (CFSE staing) and the expression of the IL-2R α-chain is shown in the absence (top) or presence (bottom) of antibodies specific for IL-2. Only cells, that express the high affinity IL-2 receptor, proliferate. Thus, in cell culture proliferation depends on the activation of the autocrine positive feedback loop.

In the previous sections, the analysis of the IL-2/IL-2 receptor gene-expression network revealed that Th cell activation depends on the activation of the autocrine positive feedback loop. The strength of the antigen stimulus has been identified as an important control parameter. In addition, due to this regulation, Th cell activation occurs in a switch-like manner. In the next section the interaction of the IL-2/IL-2 receptor gene-expression network in a system of two coupled T cells is studied.

3.6   Paracrine IL-2 signaling

As other cytokines IL-2 is a soluble mediator for cell communication in the immune system. From this perspective it is surprising that it acts in an

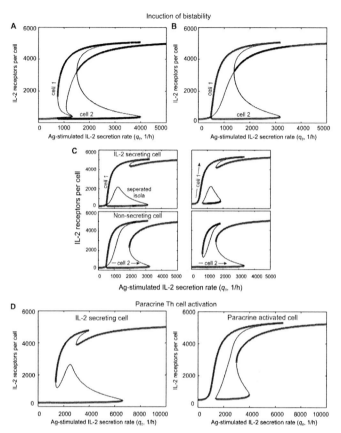

Figure 3.9. Th cell interaction (A) A pitchfork bifurcation occur in addition to the saddle-node bifurcations if two cells are coupled (stable steady state thick line, unstable steady state thin line). For small and moderate strength of antigen stimulation the activation of cell 2 is suppressed ($k_d = 10$ h^{-1}). For strong antigen stimulation both cells are activated. (B) In the case of practically no extracellular degradation ($k_d = 0.1$ h^{-1}) the coupling of two identical, interacting Th cells leads to the induction of bistability in one cell (cell2). Its activation is suppressed for small antigen-induced IL-2 secret on rates. (C) Isolated solutions occur if an advantage in either the IL-2 secretion rate (left) or the IL-2R expression (right) is assumed. (D) The situation of a IL-2 secreting Th cell (left figure) and a non-secreting Th cell (right figure) is presented. For the non-secreting Th cell an advantage in antigen-induced IL-2 receptor expression is assumed ($v_{02} = 300$ molecules/h). The non-secreting cell efficiently competes for IL-2 and upregulates its receptor in a almost gradual manner, whereas the activation of the IL-2 secreting cell is suppressed ($k_d = 0.1$ h^{-1}).

autocrine manner. The question arises if IL-2 acts only in an autocrine or also in a paracrine manner. In the following this question is addressed by studying the interaction between T helper lymphocytes. The model introduced in section 2.1 and Figure 2.2 is used. Coupling two Th cells is the simplest way to analyze Th cell interaction, but it can also account for a situation, in which each Th cell is surrounded entirely by identical cells.

First the effect on the bistable behavior of Th cell activation is studied. For this purpose the response of two identical IL-2 secreting cells to different strength of antigen stimulation is analyzed at a small intercellular distance of 10 µm. In Figure 3.9 in addition to the saddle-note bifurcation a supercritical pitchfork bifurcation emerges due to the symmetry in the system (identical IL-2 receptor dynamics and the planar symmetry in space). In a pitchfork bifurcation, the stable steady states appear and disappear in pairs. Although both cells secrete IL-2, only one cell upregulates its IL-2 receptors for small and intermediate IL-2 secretion rates (cell 1), whereas the upregulation of the IL-2 receptor on the second cell is suppressed (cell 2). Only for a strong antigen stimulus both cells are activated. An initial advantage in IL-2 receptors determines which cell is activated first.

In Figure 3.9B, one cell upregulates its IL-2 receptors gradually (cell 1) and induces bistable behavior in the other cell (cell 2), which leads to a suppressed state. Again, due to an initial advantage in IL-2 receptor numbers one cell efficiently competes for IL-2 if the extracellular IL-2 is low. This situation reveals that the strong IL-2-uptake potential of the IL-2 receptor dynamics is able to induce all-or-nothing Th cell activation in a cell with an initial disadvantage in the amount of IL-2 receptors.

In living cells, total symmetry is unlikely to occur. In the Figure 3.9C, the bifurcation diagrams for asymmetry in the IL-2 secretion or in the IL-2 receptor dynamics are shown (for a small loss of IL-2, cf. Figure 3.9B). The asymmetry leads to the formation of isolated solutions. Always the cell with an advantage in either IL-2 secretion or IL-2 receptor expression is activated first. From this observation the question arises if the inability to secrete IL-2 can be compensated by an advantage in the IL-2 receptor expression. This would represent the situation of paracrine Th cell activation. In Figure 3.9D, the response of a IL-2 secreting cell (left figure) and a non-secreting cell (right figure) to increasing strength of antigenic stimulation is shown. A small advantage in the basal IL-2 receptor expression rate is assumed for the non-secreting cell ($v_{02} = 300$ molecules/h). Indeed, the non-secreting cell deprives IL-2 from the secreting cell to activate its positive feedback loop and upregulates its IL-2 receptor expression. (In the model this results from a mergence of the two isolated solutions (Figure 3.9C), which arise if asymmetry is introduced to the system.) Therefore, an advantage in IL-2 receptor expression, even a small one, enables the Th cell to compete for IL-2. Under this condition paracrine Th cell activation occurs. Thus, the strong potential of the IL-2 receptor dynamics to take up IL-2 enables paracrine Th cell activation. Moreover, this strong competitive potential can induce switch-like upregulation of the IL-2 receptor. Pandiyan et al. [58] showed that providing Th cells with an advantage in IL-2 receptor expression enables them to compete for IL-2. However, this findings are restricted to very limited IL-2 supply.

3.7 Discussion

The IL-2/IL-2 receptor gene-expression network analyzed in this chapter exhibits two properties important for cellular interaction: (1) High IL-2 concentration is localized to the proximity of the plasma membrane of the IL-2 secreting cell due to strong re-uptake of the cytokine. (2) Th cell activation occurs in a switch-like manner. Both properties are mediated by the autocrine positive feedback of IL-2 and IL-2 receptor expression. In cell culture experiments it could clearly be shown that Th cell activation is bimodal and that proliferation is linked to high IL-2 receptor expression.

Autocrine positive feedback regulation of the IL-2/IL-2 receptor expression network Due to the autocrine positive feedback activation, the expression of the IL-2 receptor exhibits bistable behavior. This bistable behavior introduces a threshold in the antigen-induced IL-2 secretion rate, which has to be exceeded to obtain Th cell activation. Therefore, three different kinetics of the IL-2 receptor are possible depending on the strength of the antigen-induced IL-2 secretion rate: (1) For very low secretion rates no Th cell activation is observed at all. (2) For IL-2 secretion rate just above the threshold value, IL-2 has to accumulate in the extracellular medium before the autocrine positive feedback loop is activated. (3) In response to strong antigen stimulus the autocrine positive feedback is activated instantly following antigen stimulation. All this different behaviors are due to the positive feedback regulation of Th cell activation. Positive feedback loops are common regulatory motives in signal transduction and gene-expression. Positive feedback regulation is often associated with the conversion of a gradual activation signal into an all-or-nothing decision [3, 16, 73] or even with the formation of memory. In the latter case the hysteresis is so pronounced that deactivation is impossible [35, 90].

Th cell activation is controlled by positive feedback loops on multiple levels. In addition to the positive feedback loops of the IL-2 and IL-2 receptor expression, analyzed in this chapter, positive feedback regulation is also reported as part of the signaling cascades downstream of the TCR [1, 16]. In a theoretical study Brandman et al. [10] analyzed interlinked positive feedback loops. They found that the combination of a fast and a slow positive feedback loop result in resistance to noise and allows rapid activation. The sequential activation of TCR and IL-2 receptor signaling on a fast and a slow time scale allows noise filtering by the TCR signaling and leads to an integrated final decision on Th cell activation if the positive feedback of the IL-2 receptor expression is activated. Since this feedback loop is autocrine the decision is made according to the strength of the antigen stimulus and the environmental conditions. Interrupting the autocrine positive feedback by using antibodies for IL-2 (Figure 3.8C and D) prevents the upregulation of the IL-2 receptor and subsequent proliferation. Thus, the strength of the antigen stimulus not only regulates Th cell activation in all-or nothing manner, but also controls the responsiveness to external signals. Furthermore, the small shape of hysteresis (Figure 3.3A) reveals a strong coupling of IL-2 and the IL-2 receptor expression. Therefore, no memory is established by this network.

Autocrine and paracrine IL-2 signaling   Experimental measurements have shown that maximal 25% of the T cells secret IL-2, whereas up to 90% of the cells up-regulate the IL-2R α-chain (cf. Figure 1.1 and Figure 3.2). There are two possible explanations for this discrepancy: (1) every cell produces IL-2 for a short period so that at a given time point only 25% producers are detected or (2) only up to 25% of a Th cell population produce IL-2 for a extended period. Although single cell kinetics of IL-2 secretion will be required to ultimately distinguish between these alternatives, the analysis of the dynamics of the IL-2/IL-2 receptor expression network and the interaction of T helper cells support the latter possibility. (1) The strong potential of the IL-2 receptor dynamics to take up IL-2 enables paracrine Th cell activation (cf. Figure 3.9). (2) The small hysteresis of the IL-2 receptor expression in dependence of the antigen-induced IL-2 secretion rate argues in favor of prolonged high IL-2 receptor expression to obtain the observed high, sustained IL-2 receptor expression (3.1A, cf. also Appendix Figure C.1B).

During an immune response to a pathogen, only a few naive or memory Th cells of a given antigen specificity are present at the beginning of an immune response. Autocrine Th cell activation is likely to occur due to the strong re-uptake of IL-2. Moreover, it has been reported that IL-2 and IL-2 receptor expression is directed into the interface of the Th cell and the antigen-presenting cell [52]. Upon strong antigen stimulation, Th cells and the APC form this stable interaction, called the immunological synapse. Therefore, autocrine activation can be assumed for the initiation of an immune response. During the ongoing of an immune response the Th cells proliferate rapidly. The stable interaction between a Th cell and an APC is likely to be broken. Therefore, at later stages, paracrine activation could occur in addition to autocrine activation, when a sufficient number of Th cells is concentrated in the restricted space of the lymph node. The high potential of the Th cell to take up IL-2 ensures a optimal utilization of the extracellular IL-2. In the situation of a chronic infection and an autoimmune disease ,this strong up-take potential of IL-2 could emerge to be harmful. High numbers of activated Th cells are present, but the antigen stimulus can assumed to be rather weak due to a low concentration of pathogens that survived the acute immune response or due to a weak response to self-antigen. In this case the optimal up-take of IL-2 could result in an unwanted weak but continuous ongoing of the immune response. Therefore, the IL-2/IL-2 receptor gene-expression network provides the ability for autocrine and paracrine IL-2 signaling, and both modes of activation might occur *in vivo*.

Outlook The cell culture experiments have clearly shown that the IL-2 receptor expression in a Th cell population is bimodal. Th cells express either no or high amounts of the IL-2 receptor. Although the mathematical analysis argues in favor of all-or-nothing activation of the IL-2 receptor also on single cell level, an induction of this switch-like behavior as shown for the special case of very limited IL-2 cannot be excluded. An experimental validation on single cell level would be of interest. A dose-response curve of IL-2R α-chain gene-expression to different extracellular IL-2 concentration would be a first approach. The amount of cooperativity or hysteresis in the IL-2R α-chain gene-expression could be determined by means of a GFP-tagged gene. The extracellular IL-2 concentration could be controlled by using Th cells isolated from mice that are transgenic for the human IL-2R α-chain, since this receptor can not bind mouse IL-2 [17].

Chapter 4

The interaction of helper and regulatory T lymphocytes mediated by IL-2

The analysis of autocrine and paracrine uptake of IL-2 reveals the competitive potential of the IL-2 receptor dynamics due to positive feedback regulation. The cell, which has an advantage in IL-2 receptor expression, efficiently competes for IL-2. Due to the differential regulation of the transcription factor FOXP3 in Th cells and in regulatory T cells, Treg cells constitutively express high amounts of the IL-2 receptor and its expression is further up-regulated upon antigenic stimulation [17]. Because Treg cells cannot produce IL-2, they depend on paracrine IL-2 signaling. Considering the pronounced advantage of high affinity IL-2 receptor expression of regulatory T cells one can hypothesize that they may be efficient competitors for IL-2. In this chapter the interaction of Th cells and regulatory T cells is analyzed with respect to the competitive potential of the regulatory T cells for IL-2. The analysis is based on the model describing the interaction of a Th cell and a Treg cell introduced in chapter 2.1. This model accounts for a situation in which both cell types are present in the same proportion. The theoretical results are validated by cell culture experiments.

4.1 Paracrine IL-2 uptake by Treg cells shifts the Th-response threshold

As in the previous chapter, a steady-state analysis is used to depict the state of activation of both cell types for different strength of antigenic stimula-

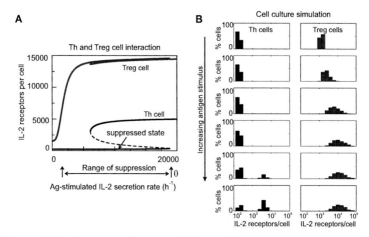

Figure 4.1. Response diagram of interacting Th and Treg cells. (A) The bifurcation diagram was computed for a proximal Th and Treg cell pair ($L = 10$ µm). Owing to the presence of the Treg, the bistability of Th-cell dynamics is enhanced and the activation threshold (θ) is strongly increased (black line) as compared to an isolated Th cell (cf. Figure 3.3). The upregulation of the IL-2 receptor is suppressed up to very high antigen-induced IL-2 secretion rates (q_0). IL-2 uptake by the Treg cell leads to a continuous up-regulation of its high-affinity IL-2 receptors (red line). (B) Computed expression of IL-2 receptor in a population of 1000 Th cells and 1000 Treg cells with log-normally distributed antigen- and feedback-driven IL-2R α-chain expression rate constants ($v_{00} = 150\pm23$ molecules/cell/h and $v_{01} = 3000\pm762$ molecules/cell/h). The distributions were computed for six increasing levels of antigen-induced IL-2 secretion rates, corresponding to increasing antigen stimulus: $q_0 = 0, 300, 600, 1800, 2700, 15000$ molecules/cell/h. The discrete switching in a single cell (cf. Figure 2C) translates into a binary response pattern of the cell population.

tion. The bifurcation diagram reveals a pronounced shift in the activation threshold (θ) of the Th cell in the presence of a regulatory T cell (Figure 4.1A, black line). Due to its advantage in IL-2 receptors, the regulatory T cell very efficiently takes up IL-2. Even for high IL-2 secretion rates ($q_0 < 20000$ molecules/h) the IL-2 receptor expression on the Th cell remains at the basal state. The paracrine consumption of IL-2 strongly enhances the IL-2R α-chain expression on the Treg cell (Figure 4.1A, red line). Although there is a region of IL-2 secretion where the increase of the number of IL-2 receptors on the Treg cell is rather steep, the entire stimulus-response curve appears continuous, unlike the switch-like activation of IL-2R α-chain expression in Th cells. Therefore, Th cell activation is prevented over a large range of antigenic stimulation due to efficient competition of IL-2 by Treg cells. The suppression can be overcome by very strong antigenic stimulation ($q_0 > 20000$ molecules/h). Therefore, competition is sufficient enough to withdraw IL-2 from the IL-2 secreting Th cell and interrupts the autocrine positive feedback loop; this results in a suppression of the upregulation of the IL-2 receptor and subsequent in the prevention of Th cell activation.

The mathematical model makes specific predictions: The system dynamics of competing positive feedback loops of IL-2 consumption by Treg and Th cells generate (1) Th-cell activation is binary as in the case without IL-2 competition; (2) the presence of Treg cells shifts the activation threshold of the Th cells to higher values of antigen dose; (3) in contrast to the binary response of the Th cells, the Treg cells utilize IL-2 in a continuous fashion, which enables the Treg cell to efficiently adapt to the IL-2 supply. For better comparison of the theoretical predictions with cell culture experiments, the IL-2 receptor expression is calculated for an ensemble of 1000 cells, again assuming a lognormal distribution of the IL-2 receptor expression rates around the mean given in Table 2.1 (cf. Figure 3.3). The difference in the all-or-nothing upregulation of the IL-2 receptor on Th cells and the gradual upregulation on Treg cells with increasing IL-2 secretion rates is clearly seen (Figure 4.1B).

4.2 Competitive IL-2 uptake of Treg cells adapts to IL-2 supply

To test these three predictions formulated in the previous section, a co-culture with primary murine Th and Treg cells from transgenic mice was prepared (cf. Materials and Methods). To control their antigen activation levels individually, Th and Treg cells were chosen to be responsive to different antigens. In separate experiments, the co-cultures were stimulated with different doses

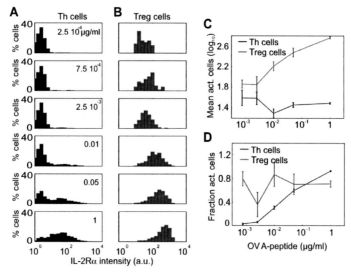

Figure 4.2. IL-2 uptake by Treg cells (A and B) In a co-culture of Th and Treg cells, the expression of IL-2R α-chain in Treg cells adapts in a graded manner to the Th-cell stimulus whereas IL-2R α-chain expression on Th cells is digital. Aggrecan-TCR$^{tg/tg}$ CD4$^+$CD25$^-$ Treg cells were co-cultured with OVA-TCR$^{tg/tg}$ CD4$^+$CD25$^-$ Th cells at 1:2 ratio. Cultures with different concentrations of OVA peptide as indicated were prepared with constant aggrecan = 2 µg/ml. Surface IL 2R α-chain determined by flow cytometry after 72 h of culture. In the activated Th and Treg cell populations the mean IL-2Rα expression (C) and the fraction of active cells (D) are characteristic of digital (blue data) and graded (red data) regulation, respectively.

of Th-cell antigen (OVA) while the Treg antigen (aggrecan) was kept constant. The expression patterns of IL-2R α-chain on Th and Treg cells were qualitatively different. The Treg cells exhibited the predicted gradual pattern of IL-2R α-chain up-regulation (Figure 4.2B). In contrast, the Th cells again showed two distinct levels of IL-2R α-chain expression in accordance with the bistable dynamics of the model (Figure 4.2A). Although the variance of the IL-2R α-chain distribution in the active state is larger in co-culture than in pure Th-cell culture and the mean value is somewhat lower (cf. Figure 3.7), the bimodal expression pattern can be clearly distinguished from the gradual behavior of IL-2R α-chain on Treg cells. To quantify this difference in IL-2 receptor upregulation of Th cells and Treg cells the mean and the fraction of the activated cells were analyzed. The IL-2R α-chain distributions on Th and Treg cells were approximated by the sum of two log-normal distributions characterizing, respectively, cells with basal IL-2R α-chain expression (non-activated cells) and cells with elevated IL-2R α-chain expression (activated cells). For Th cells, a constant mean IL-2R α-chain expression of the activated cells (Figure 4.2C, blue data) was found and a continuous, near-linear increase of the activated cell fraction with antigen stimulus (Figure 4.2D, blue data), precisely as expected for a digital gene induction. Conversely, for the Treg cells increasing mean of IL-2R α-chain expression (Figure 4.2C, red data) without a significant change in the activated fraction (Fig 4.2D, red data) was obtained, as is characteristic of a unimodal, gradual expression pattern. As in the case of pure Th-cell culture, proliferation of the Th cells correlated strongly with the activation of IL-2R α-chain expression (Figure4.3A). The antigen dose required for activation of Th cells was increased compared to the pure Th-cell culture (cf. Figure 3.8), as seen by plotting the percentage of IL-2R α-chain-positive Th cells versus the antigen dose for both cases (Figure 4.3B). The reduction in the number of IL-2R α-chain positive cells in co-culture is in accordance to the described shift of the activation threshold (cf. Figure 4.1A) and previous experimental findings [6, 17]. In addition, the proliferation of Th cells in co-culture with Treg cells was reduced for moderate strength of antigen stimulation (Figure 4.3C) emphasizing the dependence of Th cell activation and proliferation. In summary, the experimental data show the predicted divergent modes of IL-2R α-chain regulation in Treg cells (gradual IL-2 uptake) and Th cells (switch-like activation). Owing to high constitutive IL-2R α-chain expression and its enhancement through positive feedback, Treg cells can continuously adapt their IL-2 uptake capacity to the IL-2 supply and, hence, consume IL-2 more efficiently than IL-2 producing Th cells. As a result Th cell activation and proliferation is suppressed for moderate antigen dose.

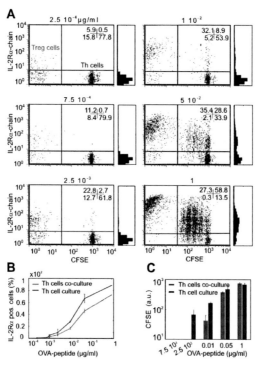

Figure 4.3. **Th and Treg cell interaction** (A) Treg cells suppress proliferation of Th cells at intermediate doses of Th-cell antigen. In the experiments described in Figure 4.2, the Th cells were labeled with CFSE; divided cells show loss of CFSE intensity, correlating with elevated IL-2R α-chain expression. The Treg cells can be identified by absence of the CFSE label. The onset of IL-2R α-chain upregulation (B) and cell proliferation (C) are shifted to higher antigen stimuli by the presence of Treg cells, while a large stimulus (1 µg/ml OVA) overcomes the inhibitory Treg effect on the proliferation of the Th cells. (Data were kindly provided by Alexander Scheffold, DRFZ, Berlin).

4.3 Distance dependence of paracrine IL-2 up-take

The result that efficient competition for IL-2 by the regulatory T cells governs cell fate decision in Th cells implicates highly regulated IL-2 action. As shown in the previous section the immunosupressive action of IL-2 is mediated by its efficient uptake by Treg cells, but how can Th cells escape competition and utilize IL-2 to serve their immunostimulatory function? Since IL-2 action is highly localized, the question arises over which spatial range competition occurs? The critical distance at which both cells receive half of the IL-2 is approximately given by

$$L_{1/2} = \frac{D}{k_{on}} \left(\frac{1}{r_1} - \frac{1}{r_2} \right), \; r_2 > r_1 \qquad (4.1)$$

(Appendix D) where r_1 and r_2 denote the IL-2R α-chain surface densities of Treg and Th cells, respectively; D is the IL-2 diffusion coefficient and k_{on} the binding rate constant of IL-2 to the high-affinity IL-2 receptor. For reasonable parameter estimates (10^4 high-affinity IL-2 receptor on Treg cells, 10^3 on an activated Th cell, cell diameter 10 µm, for values of D and k_{on} see Table 2.1), the paracrine signaling distance is 60 µm. If the actual distance between Treg and Th cell is lower ($L < L_{1/2}$), paracrine IL-2 uptake by the Treg cell will dominate; conversely, if it is larger ($L > L_{1/2}$), the IL-2 is mainly taken up by the Th cell. Therefore, the Treg cell must be located sufficiently close to the IL-2 source in order to capture the cytokine; otherwise autocrine re-uptake by the secreting cell will predominate. As a necessary condition for competition, the Treg cell must have more IL-2 receptors than the Th cell (otherwise there is no positive sharing distance). We have shown above that Treg cells may dynamically maintain their advantage in IL-2R α-chain expression even when the Th cells become activated (cf. Figure 4.1A). In addition, as $r_2 \gg r_1$ the sharing distance $L_{1/2}$ is set mainly by the autocrine IL-2 uptake capacity of the Th cell.

$$L_{1/2} \approx \frac{D}{k_{on}} \frac{1}{r_1} \qquad (4.2)$$

This explains the strong hysteresis in the Th-cell response curve (cf. Figure 4.1A). Prior to antigenic stimulation, the IL-2 uptake capacity of the Th cell is very low and the Treg cell can prevent the autocrine up-regulation of IL-2R α-chain. Once sufficient high-affinity IL-2 receptors are expressed on the Th

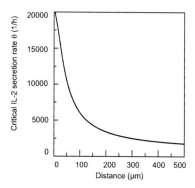

Figure 4.4. Distance dependence of the activation threshold In co-culture the activation threshold of the Th cell strongly depends on the distance. For small intercellular distances, the threshold is very sensitive to changes.

cell, however, the Treg cell cannot interrupt the now efficient autocrine IL-2 signaling. To look at the distance dependence of the activation threshold more closely, the activation threshold is plotted in dependence of the strength of the antigen stimulus and the distance (Figure 4.4). For small distances, which can be assumed to occur *in vivo*, the activation threshold is high and very sensitive to changes in the intercellular distance.

The high sensitivity of the threshold for Th cell activation to changes in the intercellular distance at small distances is very difficult to verify experimentally. But by separating Th cells and Treg cells using transwell cultures, the distance dependence of competition can be surveyed on a large scale. In transwell cultures the Th cell and the Treg cell population is separated by a membrane permeable for diffusible molecules like IL-2. In contrast to co-culture in this experimental set up Treg cells are not able to suppress T cell activation and proliferation anymore [81, 82]. The two experimental situations are illustrated as a cartoon and the related, calculated responses of Th cells and Treg cells to different strength of antigen stimulation are shown in Figure 4.5. Since the separation by a membrane can be assumed to be wide, an intercellular distance of 1000 μm was chosen. Separation of the two cell types by 1000 μm appears to be very similar to the situation in which no regulatory T cell is present (cf. Figure 3.3). Therefore, large separation of Th and Treg cells abrogate competition for IL-2, which is consistent with the model. Originally transwell culture experiments have been performed to rule out competition for soluble factors as suppressive mechanism of Th cell activation by Treg cells, but as discussed above, they rather reveal that

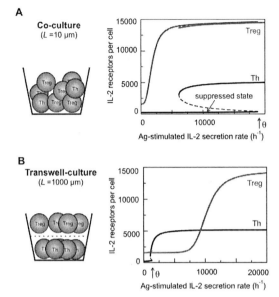

Figure 4.5. **Separation of Th and Treg cell populations in Transwell culture abrogates suppression of Th cell activation** (A) In co-culture experiments the interaction of Th cells and Treg cells is short ranged. Treg cells can suppress Th cell activation and proliferation ($L = 10$ µm). (B) In Transwell culture the Th and Treg cell population are separated by a membrane permeable for e.g. IL-2. Suppression is abrogated. The model analysis reveals that long distance ($L = 1000$ µm) between Th cell and Treg cells diminish the influence of the latter (cf. Figure 3.3).

competition is restricted to small distances.

The distance between Treg cells and Th cells can also be altered by the ratio of Th to Treg cells. The dependence of competition for IL-2 on the intercellular distance predicts a reduction in the shift of the activation threshold if the number of Th cells is increased. This corresponds to a decreased potential of Treg cells to suppress Th cell activation in cell culture. In contrast an increase in IL-2 receptor number on Treg cells, for example by exposure to IL-2 previous to co-culture with Th cells, should increase their suppressive potential. Both situations can be accounted by mathematical modeling.

A model is constructed describing the interaction of two Th cells and one Treg cell (Figure 4.6). Placing the Treg cell in the center of the one dimensional

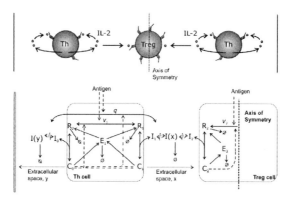

Figure 4.6.   **Interaction of Th cells to Treg cells at a ratio of 2:1**
Model symmetry is used to reduce the complexity of the model. The IL-2 and
IL-2 receptor dynamics is modeled as described in Figure 2.1. Additionally the
feedback-dependent IL-2 receptor expression integrates over the number of IL-2/IL-
2 receptor-complexes at the plasma membrane facing the Treg cells and directing
away from it. The diffusion at the boundary of the system is limited by a barrier
as described in Figure 2.2.

arrangement of cells, one can exploit the symmetry of the system. Diffusion in
the space directed away from the Treg cell is limited by a diffusion barrier as
described in Figure 2.2. In Figure 4.6, a scheme of the situation is presented,
whereas the model equations are given in detail in Appendix B.3. IL-2 is
secreted to equal quantities in the extracellular space facing towards or away
from the Treg cell. It is assumed that the Th cell can integrate over the IL-2
receptors on its surface leading to a reduction in the activation threshold
by approximately half for small distances (Figure 4.7, compare solid and
dashed line). Therefore, the reduced ability of the Treg cell to deprive the
Th cell from the IL-2 can be followed by the mathematical model and was
also observed in cell culture experiments [17].

Preactivation of Treg cells by stimulating them in the presence of IL-2 previous to co-culture greatly enhances their suppressive capacity [17]. One can
hypothesize that this is due to its increased potential to compete for IL-2
due to their elevated numbers of IL-2 receptors expressed. In the model,
an increase in the basal IL-2 receptor expression rate by 8-fold reverses the
reduction of the activation threshold reaching approximately the situation
1:1 (Figure 4.7, dot-dashed line), but it cannot account for the elevated suppressive potential of pre-activated Treg cells as observed in cell culture [17].
Other mechanisms than competition for IL-2 may account for this difference.

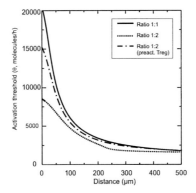

Figure 4.7. Competition depend on the ratio between Th and Treg cells The Treg cell increases the activation threshold of the Th cell through IL-2 deprivation if it is located in proximity (solid line, cf. Figure 4.4). The effect of IL-2 competition is also seen for a Th:Treg ratio of 1:2 but is diminished (dashed line). IL-2 competition can be restored when Treg cells are preactivated, giving them elevated IL-2R α-chain levels already at the beginning of co-culture with the Th cells (dot-dashed line).

However, the qualitative agreement of the theoretical and experimental findings suggests that efficient competition for IL-2 also depends on the ratio of Th cells and Treg cells.

4.4 Regulation of Th cell activation

The upregulation of the high affinity IL-2 receptor, which is referred to as Th cell activation, can be altered by efficient competition for IL-2 by Treg cells. The properties that regulate the interaction between Th cells and Treg cells have been determined in the previous chapters and the described mechanisms for competition were validated experimentally: (1) Bistable expression pattern of the IL-2 receptor expression in Th cells leads to all-or-nothing activation; (2) In contrast, Treg cells adapt their IL-2 receptor expression gradually to the IL-2 supply making them efficient competitors for IL-2; (3) Due to the high uptake rate of the IL-2 receptor, IL-2 action and subsequently competition for IL-2 is very localized. Therefore, two control parameters are identified that regulate the activation state of Th and Treg cells. The strength of the antigen stimulus regulates the amount of IL-2 secreted and is therefore the key parameter for Th and Treg cell interaction. Since IL-2 ac-

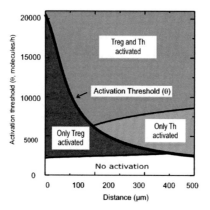

Figure 4.8. Outcome of competition Phase diagram for the behavior of a
Th-Treg cell pair in dependence on IL-2 secretion rate by the Th cells (q) and
intercellular distance (L). The locus of the activation threshold for Th cells (θ,
thick black line) and the transition line indicating activated Treg cells (thin black
line, chosen at a characteristic value of 10000 up-regulated IL-2Rs per cell) divide
Th parameter plane in future regions. The different regions indicate whether IL-2R
α-chain expression is activated by positive feedback and thus the cells receive IL-2
signal. All possible combinations of activations can occur as a function of the two
control parameters L and q: only Treg cells activated (red region), only Th cells
activated (blue region), no cell activated (white region), and both cells activated
(gray region).

tion is localized, the distance is the other control parameter. In dependence
of these two parameters distinct activation pattern of the two cell types can
be defined (Figure 4.8). For very low antigen IL-2 secretion rates, there is
no activation at all (white region). For small distances, Th cell activation is
suppressed up to high IL-2 secretion rates and therefore only Treg cells are
activated (red region). For weak or intermediate antigen stimuli and a high
distance, only the Th cells are activated (blue region). Increasing the antigen
stimulus results in the activation of both cell types (gray region). Thus, the
strength of the antigen stimulus and the intercellular distance are identified
as two important control parameters for Th and Treg cell interaction. The
resulting distinct activation pattern may facilitate the understanding how
IL-2 can serve immunostimulatory and immunosuppressive functions.

4.5 Discussion

In this chapter, we have identified the mechanisms and control parameters determining the outcome of competition for IL-2 between Treg and Th cells. The gradual upregulation of the IL-2 receptor enables Treg cells to adapt to the extracellular IL-2 supply. This mechanism combined with the advantage in IL-2 receptor expression makes Treg cells very efficient competitors for IL-2. As a consequence of this efficient competition the activation threshold of Th cells is increased and very strong antigenic stimulation is required for their activation. The outcome of competition is determined by the strength of the antigen stimulation, the intercellular distance and also the ratio of competing Th and Treg cells.

Mechanism of competition for IL-2   In both cell types, Th and Treg cells, the upregulation of the IL-2 receptor is controlled by positive feedback regulation. In Th cells this positive feedback regulation results in an all-or-nothing activation. In contrast to Th cells, Treg cells gradually increase their number of IL-2 receptors. Due to the continuously high IL-2 receptor expression in Treg cells, mediated by the transcription factor FOXP3, no activation threshold has to be overcome. The positive feedback regulation rather acts as on amplifier of IL-2R α-chain expression. This difference in IL-2 receptor upregulation was clearly revealed by cell culture experiments. The experiments show that the increase in the amount of activated Th cell appears to be bimodal for increasing strength of antigenic stimulation whereas it is gradual in Treg cells (cf. Figure 4.2). Gradual upregulation of the high-affinity IL-2 receptors allows perfect adaptation to the extracellular IL-2 supply. Adaptation therefore provides an efficient mechanism to compete for IL-2 (cf. Figure 4.1A). In the mathematical model Treg cells can only adapt to the IL-2 supply for relatively small IL-2 secretion rates. In the cell culture experiment, however, Treg cells upregulate their IL-2 receptor expression continuously in response to a wide range of antigen stimuli. This additional capacity of adaptation to external IL-2 supply may be due to the additional regulation of the IL-2R α-chain gene-expression by the transcription factor FOXP3, which is not included in the model.

In contrast to Th cells that have the ability to induce switch-like activation in the case of strong coupling by very limited IL-2 supply (cf. 3.9B), Treg cells are efficient 'natural' competitors due to their differential regulation of the IL-2 receptor expression. Their suppressive action of IL-2 deprivation can only be overcome by a very strong antigen stimulus.

Control parameters of competition for IL-2   The antigen stimulus is clearly the most important control parameter of competition for IL-2, since its strength determines whether or not extracellular IL-2 is a limited source. However, it was shown that Treg cells are only able to suppress Th cell activation and proliferation, if they are present in co-culture not later than 6-10 hours after antigen stimulation [78]. Therefore, a time window for competition exists. In chapter 3 (Figure 3.1), we have shown that the time until the initiation of proliferation depends on the strength of the antigen stimulus. This may be due to the fact that in the case of a weak or moderate antigen stimulus, IL-2 has to accumulate in the extracellular medium before a critical concentration for activation is reached. The requirement of interrupting the autocrine positive feedback at early stages to prevent accumulation of IL-2 in the extracellular medium would also explain the observation of a time window for competition. This remains to be tested, though.

The intercellular distance has been identified as a second control parameter. Proximity between Th and Treg cells is required for competition. In the lymph nodes, proximity can be assumed because of the small intercellular distances between neighboring cells (Figure 1.4). Moreover, it is likely that Th cells and Treg cells are neighboring cells if both have the same antigen specificity. The distance-dependence of competition was also shown in a two-dimensional arrangement of Th and Treg cells [83]. The influence of the number and the ratio of competing cells on the outcome of competition was analyzed by changing the ratio of Th and Treg cells to 2:1. *In vivo* only 5-10% of the naive T cell pool are naturally occurring Treg cells. However, several mechanisms are reported to overcome this low representation [68]. For example, it was observed that the number of Treg cells increase with the number of IL-2 secreting Th cells [2]. Further mathematical analysis in which the number of competing cells depends on efficient competition for IL-2 as observed would be of interest.

Competition for IL-2 as suppressive mechanism of Th cell activation and proliferation   Various suppressive mechanisms of Th cell activation, survival and effector function by Treg cells have been described in the last years [84]. Next to competition for IL-2 other cytokine mediated or cell-cell contact dependent mechanisms are described: (1) Treg cells secrete the inhibitory cytokines IL-10, IL-35 and TGFβ. Whereas IL-35 is required for Treg cells to exert their maximal suppressive activity, IL-10 and TGFβ are involved in the control of allergy and inflammatory diseases (e.g. colitis and airway inflammation) by Treg cells. They also limit anti-tumor immunity. (2) It was reported that cyclic AMP (cAMP) suppress Th cell function by transfer through gap junctions. (3) Suppression by cytolysis occurs mainly through the induction of apoptosis by granzyme B. (4) Several mechanisms by which Treg cells alter Th cell function by acting on dendritic cells are described. Dendritic cells are important for antigen presentation and Th cell activation in the peripheral lymphoid tissues. Altering dendritic cells function can, therefore, prevent Th cell activation. Treg cells further induce the expression of immunosuppressive molecules such as CTLA4 on dendritic cells. The downmodulation of their capacity to activate Th cells and suppression of their maturation is also reported. Competition for IL-2 is, thus, only one of multiple mechanisms Treg cells exert to suppress Th cell function. The relative importance of the described mechanisms remains to be determined. Currently all these mechanisms are regarded to be limited to certain diseases or conditions. In addition to the impairment of Th cell function, IL-2 enhances the suppressive activity of Treg cells. IL-2 signaling primes Treg cells to secrete IL-10 [17, 6]. IL-2 is also discussed to be involved in the secretion of IL-35 [84]. Moreover, IL-2 is important for Treg cell homeostasis. Therefore, competition for IL-2 can serve as a direct mechanism to suppress Th cell activation and proliferation but also mediates Treg cell survival and the induction of other suppressive mechanisms. How deprivation of IL-2 exactly affects Th cell function, remains to be elucidated. The cell culture experiments suggest that Th cell activation and proliferation is suppressed. The observed absence of activated and proliferating Th cells might, however, also be the result of the strong induction of apoptosis. Pandiyan et al. [58] addressed this question and concluded that IL-2-deprivation mediates suppression by inducing apoptosis not by suppressing of Th cell activation. This question is addressed in more detail in the outlook.

Chapter 5

Outlook: Analysis of Th cell proliferation in the presence of regulatory T cells

How Treg cells influence Th cell proliferation is still under debate. Pandiyan et al. [58] analyzed this interaction with respect to IL-2 deprivation in cell culture and *in vivo* and concluded that competition for IL-2 does not prevent Th cell activation and subsequent proliferation but induces apoptosis. In this outlook, initial results are presented to address the question if Th proliferation or survival is effected by Th and Treg cell interaction.

We addressed this question by measuring cell proliferation kinetics in cell culture and fitting the obtained data to the Smith-Martin model of cell proliferation. The Smith-Martin model of cell proliferation provides a simple but very efficient way of describing cell proliferation [74], but also includes apoptosis. In this model, the cell cycle is divided in an intermediate phase (A) which corresponds to G_1 phase of the cell cycle or parts of it and a cycling phase (B) which includes all phases of the replication (S to M phase). Cell death is included in both phases (Figure 5.1).

This concept of the cell cycle was first formulated by means of differential equations by Cain and Chao [12, 13]. The description of the cycling phase (B phase) with a delay differential equation allows to account for time period between the intermediate phases (A phase). Here, we consider the phases before the onset of proliferation (A_0 and B_0) separately. These phases include the upregulation of the high-affinity IL-2 and last until the time of first division. After the cells have entered the cell cycle, they pass through the different phases (A_n and B_n) of each generation ($n = 1, 2, ...$). Assuming a

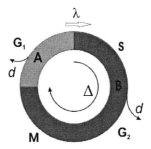

Figure 5.1. **Scheme of the Smith-Martin Model**

---

constant duration time in the B phase (Δ) [59]. The model equations read as followed

$$\frac{dA_0}{dt} = -(\lambda_0 + d_0)A_0 \tag{5.1}$$

$$B_0 = \lambda_0 \int_0^{\Delta_0} e^{-ds} A_0(t-s)ds \tag{5.2}$$

$$\frac{dA_1}{dt} = 2\lambda_0 e^{-d\Delta_0} A_0(t-\Delta_0) - (\lambda + d)A_1 \tag{5.3}$$

$$\frac{dA_n}{dt} = 2\lambda e^{-d\Delta} A_{n-1}(t-\Delta) - (\lambda + d)A_n \tag{5.4}$$

$$B_n = \lambda \int_0^{\Delta} e^{-ds} A_n(t-s)ds \tag{5.5}$$

It is assumed that at the time of stimulation no cells have entered the cell cycle ($A_0(0) = 1$ and $B_0(0) = A_n(0) = B_n(0) = 0$). The parameter λ_0 describes the rate of cell cycle entry and therefore includes Th cell activation. The rate constant d_0 accounts for cell death before initiation of proliferation. The time period after stimulation until the cells enter the cell cycle is given by Δ_0. Once the cells have entered the cell cycle, the rate constant d describes the rate of apoptosis and Δ the period of time in the B phase independent of the generations. Thus $e^{-d\Delta}$ determines the fraction of cell that survived the B-phase and $T = \frac{1}{\lambda} + \Delta$ the mean duration time of the cell cycle. This model is fitted to the relative number of Th cells in each generation over time.

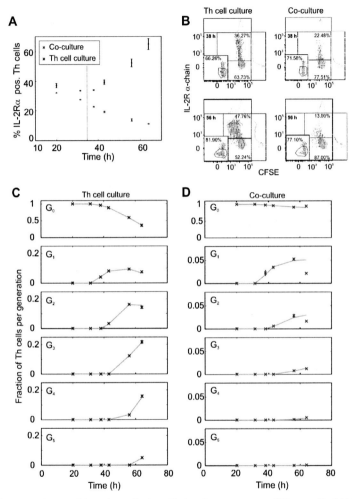

Figure 5.2. Proliferation of Th cells in the presence of Treg cells (A) Kinetics of activated Th cells in the absence (blue) or presence (red) of Treg cells *Ex vivo* isolated T cells were stimulated with 0,33 µg/ml αCD3 and irradiated APCs and IL-2R α-chain expression was measured. The amount of activated Th cells is only reduced significantly after the onset of proliferation (dotted line). (B) The amount of IL-2R α-chain expression and proliferation (by CFSE-staining) is shown for both culture conditions at 38h and 56h after stimulation. IL-2 receptor expression and proliferation are remarkably reduced in co-culture, whereas no increase in cell death could be observed (inlayed Figures). (C) and (D) Data (blue and red, cf. (A)) and best fits (gray line) were presented for the fraction of Th cells in the different generation of cell cycle (G_0-G_5). Almost no initiation of proliferation could be observed in co-culture (D).

Ex vivo isolated T cells were stimulated with anti-CD3 (0.33 µg/ml) and irradiated APCs. Th cells (CD4$^+$CD25$^-$) were culture in the presence or absence of Treg cells (CD4$^+$CD25$^+$). Th cell proliferation was followed over 3 days. At different time points cells were harvested and stained for the CD4 and CD25 (IL-2R α-chain). Live and dead cells were identified by using Dapi, a fluorescent stain that strongly binds to DNA. In Figure 5.2A the amount of activated Th cells is shown over time in the absence (blue) or presence (red) of Treg cells. This figure reveals that in the early stage after stimulation the number of activated Th cells in both culture conditions are approximately the same, only after the onset of proliferation after approximately 35 h significantly different fractions of activated Th cells are observed (cf. Figure 3.1). This reduced Th cell activation in co-culture is accompanied by a decreased expression of the IL-2R α-chain and greatly reduced but not completely abrogated proliferation of Th cell in co-culture (Figure 5.2B). Interestingly, no increased numbers of dead cells are observed in the presence of Treg cells as the inlay figures and Appendix Figure E.1 at 38 h and 56 h after stimulation. In the following the total amount of CD4+ Th cells is used for the fitting analysis. From these data we determined the relative fraction of cells in each generation by means of the proliferation marker CFSE.

The best fits obtained with the Smith-Martin model described in Eqs. 5.1-5.5 are shown in Figure 5.2C and D. This model for cell proliferation that only differentiates whether or not Th cells have entered the cell cycle can follow the data. While in the absence of Treg cells cell proliferation is observed up to generation 5 (G_5), the kinetics of the different generations (G_0-G_5) reveal that hardly any Th cells enter the cell cycle in the co-culture with Tregs (Figure 5.2D). (Note the different scales for the percentage of cells per generation in Figure 5.2C and D.) The estimated parameters with 95% confidence intervals are given in Table 5.1. The confidence intervals are obtained by profile likelihood [62]. No confidence intervals were given, if the α-quantile of the χ^2-distribution could not be reached by increasing or decreasing the parameter value by 100 times.

The parameter λ_0 describes the rate of entry into the cell cycle. The best fit values for this parameter is higher in the presence of Treg cells. However, a great uncertainty of this value is revealed by the range of the confidence intervals of the Th cells in co-culture. Hence, no significant difference between the rate constant describing the entry into the cell cycle is observed in the presence or absence of Treg cells. This result is in agreement with the observation that at early time points after stimulation no pronounced differences could be seen in the fraction of activated Th cells (Figure 5.2A). The death rate of the undivided cells also does not differ significantly, nor

	Th cell culture			Co-culture with Treg cells		
	Best fit (h⁻¹)	Confidence intervals 95% (h⁻¹)		Best fit (h⁻¹)	Confidence intervals 95% (h⁻¹)	
Aktivation rate constant (λ_0)	0.0025	0.0019	0.0034	0.0373	0.0024	0.0257
Apoptosis rate constant (d_0)	0.045	-	0.199	0.0714	0.033	0.294
Time to first division (Δ_0)	33.63	32.92	34.16	34.69	30.40	37.00
Rate constant of cell cycle entry (λ)	0.108	0.094	0.124	0.055	0.032	0.093
Apoptosis rate constant (d)	0.001	-	-	0.190	0.025	0.350
Time in B-phase (Δ)	3.996	3.669	4.375	3.045	0.795	4.677
Mean duration time of the cell cycle	13.29	11.73	15.01	21.23	11.55	36.02

Table 5.1: **Parameter values** Best fit and confidence intervals for the model parameters of the Smith-Martin model

does the difference in the time before first division or the time period the cells spend in the B phase after initiation of proliferation.

An increase in the rate of apoptosis is revealed by the best-fit values of cell death in co-culture compared to the culture without Treg cells. Unfortunately, the interpretation of the rate of apoptosis of proliferating cells (d) is impeded by the inability to define confidence intervals for this parameter describing apoptosis in the absence of Treg cells. To better understand, why these confidence intervals cannot be determined, the correlation between the apoptosis rate constant of proliferating cells and the other parameters is analyzed. For this purpose the results of the profile likelihood method are used. The confidence intervals are calculated by varying one parameter and fitting the rest to obtain the minimal χ^2 for this parameter set until a threshold value for the χ^2 is reached, which is determined by the α-quantile of the χ^2-distribution.

In Figure 5.3, the values for χ^2 and the other model parameters are plotted against the apoptosis rate constant (d) of proliferating cells. In contrast to the presence of Treg cells (Figure 5.3B, top left), the profile of the obtained χ^2 reveals no changes for the rate constant of cell death in the absence of Treg cells (Figure 5.3A, top left). Thus, the definition of confidence intervals for this parameter is impeded for d in the absence of Treg cells. Varying the apoptosis rate constant has also no influence on the model parameter except for the rate constant of cell death before entry into the cell cycle (d_0, Figure 5.3A, top right) in pure Th cell culture. Here, the correlation is strong revealing that one cannot determine these two parameters independent from each other. The same strong correlation is observed for the apoptosis rate constant before and after entry into the cell cycle for Th cells the co-culture (Figure 5.3B, top right). Thus a strong correlation exists between the death rates before and after the entry of the cell cycle. The steepness ($\Delta d_0/\Delta d$) of this function gives the ratio of the coupling. For pure Th cell culture, the steepness is approximately one, meaning that the apoptosis rate constant of the proliferating cells does not change with the entry of the cell cycle. Th ratio is approximately $1/2$ for the apoptosis rate constants of the Th cells in co-culture. This can be interpreted as an increase of apoptosis once the cells has entered the cell cycle. Thus, these data suggest an increase in cell death induced by Treg cells after the Th cells has entered the cell cycle. The only parameter value, in which the two situations differ significantly, is the rate by which the cells enter a new round of cell division (λ). This result indicates that Th cell proliferation is reduced in addition to the induction of apoptosis.

These results clearly require verification. The interpretation of the results would also be enhanced by some improvements in the acquisition of the data. Measurements of the IL-2R α-chain expression at earlier time points would allow to estimate the rate of Th cell activation. Taking the absolute cell numbers into account provides a important step to better determine cell death. The estimation of cell death directly from cell culture experiments is difficult, because apoptosis is rapidly followed by cell lysis. These lysed cell parts are not detected by FACS analysis. This fast cell lysis would also explain the discrepancy between the absence of enhanced cell death determined by FACS-analysis (Figure 5.2B) and the increased apoptosis rate revealed by the fits. Measurements of the amount of apoptotic cells cannot solve this problem, but would give a better understanding of the condition of the cells. In addition, repeating these experiments in the presence of antibodies specific for IL-2 instead of Treg cells and the addition of IL-2 to a co-culture of Th and Treg cells would ensure if the suppression of proliferation

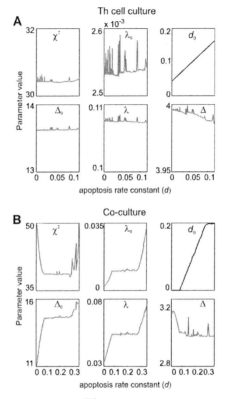

Figure 5.3

and the induction of apoptosis is an effect of IL-2 deprivation by Treg cells.

Thus, the Smith-Martin model provides a promising approach to quantify the effects on Th cell proliferation induced by Treg cells. The analysis reveals that differences between the two culture conditions only occur after the cells have entered the cell cycle. Then proliferation is impaired and the data suggest an increase in cell death. Both observation were reported before. Pandiyan et al. [58] found enhanced apoptosis of Th cells in co-culture with Treg cells. Duthoit et al. [25] reported that Treg cells induce an uncoupling of IL-2 signaling from cell cycle progression. They found that expression of cyclins and c-Myc is suppressed resulting in an arrest in the G_0-G_1-Phase.

Conclusions

In this work, we have shown that positive feedback regulation of the high-affinity IL-2 receptor expression induces different modes of its upregulation in Th and Treg cells. In Th cells, the autocrine positive feedback control leads to switch-like upregulation of this receptor, whereas Treg cells gradually adapt their upregulation of the IL-2 receptor to the extracellular IL-2 supply due to changes in IL-2 receptor expression mediated by the transcription factor FOXP3. This differential regulation of the IL-2 receptor expression in Treg cells makes them very efficient competitors for IL-2. However, the outcome of competition is controlled by the strength of the antigen stimulus and the intercellular distance. This distinct activation of Th or Treg cells depending on these parameters provides an explanation how IL-2 can mediate its immunostimulatory and immunosuppressive effects. Moreover, the strong re-uptake of IL-2 due to positive feedback regulation, activation induced IL-2 receptor internalization and recycling allows not only strong coupling between Th cells and Treg cells but also between Th cells in the case of very limited IL-2 supply. This strong coupling makes communication via IL-2 a highly deterministic and regulated process, counter-intuitive to what one would assume for the interaction via a diffusible molecule.

Competition for external sources are common in the immune system. For example competition for the survival cytokine IL-7 is discussed to control the size of the naive and memory Th cell population. IL-7 is secreted by stroma cells. Thus the cells that secrete IL-7 take not part in the competition. Another example are Th cell clones that compete for limited presented antigen in the course of an infection. Mathematical models addressing competition in the immune system mainly focus on the quantitative description of competition for an external source as described in the examples above [9, 34, 54, 4]. In this work, competition for an 'internal' source was analyzed. Th cells, which secrete IL-2, take also part in the competition. This kind of competition may not be an exclusive mechanism to regulate cell communication of Th and Treg cells mediated by IL-2, since (1) most cytokines are organized in expression networks of the cytokine itself and at least parts of its receptor, (2) autocrine positive feedbacks are common in regulation of cytokine expression. Moreover, this work shows that mathematical modeling combined with cell culture experiments can provide a powerful tool to describe regulation of cell interaction in the immune system as also revealed by the approach to quantify the effects of Treg cell interaction on Th cell proliferation. Thus, hopefully, further investigation on cell communication mediated by IL-2 or other cytokines may be stimulated by this work.

Chapter 6

Materials and Methods

Model analysis

Model analysis The software packages Mathematica (Wolfram Research) or Matlab (MathWorks) were used for the model analysis. The bifurcation studies were performed using the program package Xppaut (Bard Ermentraut, www.pitt.edu/~phase).

Determining the fraction of activated and non-activated cells from the distribution of the IL-2R α-chain expression The measured distribution of the IL-2R α-chain expression within a cell culture depend on the antigen concentration added but also on variations in the strength of antigenic stimulation and the spacial distribution of IL-2 concentration independent of the experimental controlled stimulus strength. The observed IL-2R α-chain expression for both cell types is interpreted as the sum of two distributions: One describing the IL-2R α-chain expression pattern of the non-activated cell population and the other describing the elevated IL-2R α-chain of the activated cells. The width of these two overlapping distributions depends on the expression pattern of IL-2R α-chain before stimulation, as does the mean IL-2R α-chain expression rate of the non-activated cell population. The following analysis was performed in order to determine the dependence of the fraction and mean IL-2R α-chain expression of the activated cell population on the applied stimulus. The result is shown in Figure 4.2 C and D. Assuming log-normal distributions for both populations, the analysis of the data shown in Figure 4.2 A and B lead to a parameter estimation problem of a probabilistic mixture model with five independent parameters for each stimulus: One mean value and one width for each of the two log-normal distributions and the weighting factor corresponding to the fraction of activated cells. The measured IL-2R α-chain expression is a sample drawn from the sum of the two distributions and thus the standard χ^2-method for binned data was used in the fitting procedure. The minimization of the χ^2-sum was performed using the trust-region reflective Newton method. Finally the uncertainty of the best-fit parameters was assessed by resampling the unbinned data (non-parametric bootstrap) with $n = 10000$ (The analysis was kindly provided by Michael Floßdorf, DKFZ, Heidelberg).

Data fitting   The data presented in the outlook were first fitted with Simulated Annealing [45]. The initial temperature was taken as 100 and final temperature as 0.02. As initial values, random values were generated restricted by appropriate upper and lower bounds. The one with the best χ^2 was chosen out of 100 fits and this fit was optimized by using the *lsqnonlin*-function implemented in Matlab; again using random values generated with the same restriction in the upper and lower bounds. The confidence intervals were determined by Profile Likelihood (PL). In this method, one parameter (θ_i) is varied and the remaining parameters were fitted again with the *lsqnonlin*-function ($\chi^2_{PL}(\theta_i) = \min_{\theta_{j \neq i}}[\chi^2(\theta)]$, [62]). This procedure is repeated until the threshold Δ_α was exceeded or the parameter value under estimation was increased or decreased by 100 fold. $\Delta_\alpha = \chi^2(\alpha, df)$, where α is the α quantile of the χ^2-distribution in and df the degrees of freedom. For this analysis, $\alpha = 0.05$ and $df = 5$, since 5 out of 6 parameters could be varied freely. Thus, $\Delta_\alpha = 11.07$. Confidence intervals were given in Table 5.1, if this threshold could be exceeded otherwise no confidence intervals were given. The matlab code for the fitting routine by Simulated Annealing and for the Profile Likelihood were kindly provided by Edda Schulz, Humboldt University Berlin.

Cell culture experiments

Mice   BALB/c OVA-TCRtg/tg DO.11.10 mice or wild type C57/Bl6 mice were purchased from the BfR (Berlin, Germany); BALB/c aggrecan TCRtg/wt (5/4E8) (peptide sequence: ATEGRVRVNSAYQDK) were obtained from Wilem van Eden (University of Utrecht). All mice were housed in a specific pathogen-free (SPF) environment and were used at 8 to 10 weeks of age.

Puffers and medium   PBS (8 g/l NaCL, 0.2 g/l KCl, 0.2 g/l KH_2PO_4, 1.4 g/l $Na_2HPO_4 \bullet H_2O$), PBS/BSA (PBS + 2 g/l cave serum albumin), PBA

(PBS/BSA + 0.05% NaH$_3$), erythrocyte-lysis puffer (10 mM KHC03; 155mM NH$_4$CL; 0.1 mM EDTA, pH 7.5), medium (RPMI-1640 supplemented with 10% heat inactivated FCS, 100 U/ml penicillin plus 100 U/ml streptomycin, 2 mML-glutamine and 50 lM2-ME (Sigma))

Antibodies   The following anti-mouse antibodies were either conjugated in our house or purchased as indicated: FITC- or PE-conjugated anti-CD4 (GK1.5, BD-PharMingen), allophycocyanin-conjugated (APC) anti-CD25 (PC61, BD-PharMingen), biotinylated anti-CD25 (7D4, BD-PharMingen), biotinylated anti-CD25 F(ab)2 (PC61, DRFZ), anti-CD25-PE (7D4, Miltenyi Biotec, Bergisch Gladbach, Germany), anti-PE-Cy7 (G155-178, BD-PharMingen), Cy5-conjugated anti-DO.11.10 OVA-TCR (KJ1.26), anti-CD3 (BD-PharMingen, 145-2C11).

Cell staining and purification   Spleen and lymph nodes were taken from the mice. Cell were isolated by straining and erythrocyte were lysed (3 min.). For isolation the suspended CD4+CD25+ Treg cells, CD4+CD25- Th cells were stained with biotinylated anti-CD25 F(ab)2 followed by incubation with anti-biotin microbeads and sorted by AutoMACS™ (Miltenyi Biotec). Subsequently, CD25- cells were labelled with anti-CD4 microbeads and sorted for CD4 expression. Antigen presenting cells were sorted using anti-MHC class II microbeads. The purity of the various sorted cell populations was higher than 95%. Before culturing the MHC class II+ (APCs) were irradiated (30 Gray). Cells were suspended in PBS/BSA and kept on ice during the whole preparation.

CFSE labeling   CD4+CD25- or CD4+CD25+ T cells were washed with PBS, resuspended in a 1 mM solution of carboxy-fluorescein diacetate- succinimidyl ester (CFDA-SE) (Sigma, St Louis, MO) at a density of 1x107 cells/ml and incubated for 4 min at room temperature. The labeling reaction was stopped by washing with RPMI 1640 culture medium (BioWhittaker, Walkersville, MD) containing 10% fetal calf serum (FCS).

alternative protocol: CD4+CD25- were resuspended in PBS + 10% FCS at a density of 2x107 cells/ml. 10μM CSFE-PBS of the same volume were prepared. Solutions were mixed and cells were incubated for 5 min at room temperature. The labeling reaction was stopped by washing with PBS + 10% FCS (3x).

Proliferation assays   0.33x106 irradiated APC and 0.18x106 T cells in
total were incubated for 72 h in a 96-well U-bottom plate. CD25+ Treg cells
and CFDA-SE-labeled CD25– Th cells were mixed in a 1:2 ratio or cultured
alone. Treg cells from aggrecan TCRtg/wt mice were stimulated with 2
µg/ml aggrecan peptide (Agg70-84), the CD4+CD25- OVA-TCRtg/tg Th
cells as indicated: 0, 0.00025, 0.00075, 0.001, 0.0025, 0.01, 0.05, 1 µg/ml
Ova323-339-peptide. RPMI-1640 supplemented with 10% heat inactivated
FCS, 100 U/ml penicillin plus 100 U/ml streptomycin, 2 mML-glutamine
and 50 lM2-ME (Sigma) was used for cell cultures.

alternative protocol: 0.2x106 T cells in total were stimulated with 0.4x106
irradiated APC and 0.33 µg/ml anti-CD3 and incubated for 20 h, 31.5 h
38 h, 43 h, 56 h and 64 h in a 96-well U-bottom plate. CD25+ Treg cells
and CFSE-labeled CD25– Th cells were mixed in a 1:1 ratio or cultured
alone. RPMI-1640 supplemented with 10% heat inactivated FCS, 100 U/ml
penicillin plus 100 U/ml streptomycin, 2 mML-glutamine and 50 lM2-ME
(Sigma) was used for cell cultures.

FACS analysis   After cell culture, cells were harvested and stained for
CD4, CD25 and Ova-TCR (anti-Ova-TCR-Cy5) for subsequent analysis of
CD25 expression. Dead cells were excluded via counter staining either with
propidium (PI, Sigma) or Dapi (Sigma). Samples were measured with FACS-
Calibur (BD) or with LSRII (BD). Acquired cells were analyzed using FlowJo
(© Tree Star, Inc. 1997-2006).

The main protocol is also published in [17].

Appendix A

Approximation of the relaxation time

In order to approximate the time, that is required to reach the steady-state distribution of the IL-2 concentration in the extracellular medium, a simplified model is considered neglecting the IL-2 receptor dynamics. The extracellular IL-2 concentration (I) is given by the partial differential equation

$$\frac{\partial I}{\partial t} = D\frac{\partial^2 I}{\partial x^2}. \tag{A.1}$$

At the boundary $x = 0$, a cell which secretes IL-2 with the rate q and takes up IL-2 with the rate constant k_1 is assumed. Whereas the cell, located at $x = L$, only takes up IL-2 with the rate constant k_2. The boundary conditions read as follows:

$$-D\frac{\partial I}{\partial x}\bigg|_{x=0} = q - k_1 I \tag{A.2}$$

$$-D\frac{\partial I}{\partial x}\bigg|_{x=L} = k_2 I \tag{A.3}$$

The constants k_1 and k_2 include the association rate constant of IL-2 to its receptor k_{on} and the number of IL-2 receptor on either cell, therefore $k_1 = k_{on}R_1$ and $k_2 = k_{on}R_2$ (cf. Figure 2.2). The initial condition is given by $I(x,0) = 0$. The solution of this partial differential equation can be expressed as the sum of the steady-state solution I_{ss} and a part describing the relaxation into the steady state $\tilde{I}$ [28]. The IL-2 concentration in the extracellular medium (I) is given by

$$I(x,t) = I_{ss}(x) + \tilde{I}(x,t). \tag{A.4}$$

$\tilde{I}$ is given by the partial differential equation with homogeneous boundary conditions and the initial condition:

$$\frac{\partial \tilde{I}}{\partial t} = D\frac{\partial^2 \tilde{I}}{\partial x^2} \tag{A.5}$$

$$-D\frac{\partial \tilde{I}}{\partial x}\bigg|_{x=0} = -k_1\tilde{I} \tag{A.6}$$

$$-D\frac{\partial \tilde{I}}{\partial x}\bigg|_{x=L} = k_2\tilde{I} \tag{A.7}$$

$$\tilde{I}(x,0) = -I_{ss} \tag{A.8}$$

This parabolic differential equation can be solved by separation of the variables. The general solution is given by

$$\tilde{I}(x,t) = e^{-D\lambda^2 t}\left[c_1\sin(\lambda x) + c_2\cos(\lambda x)\right]. \tag{A.9}$$

The relaxation of the IL-2 concentration to the steady-state distribution in the extracellular medium is given by λ, which is called the eigenvalue of the boundary-value problem. The eigenvalues λ_n $(n = 1, 2, 3...)$ have to satisfy the following equation

$$\tan(\lambda_n L) = \frac{D\lambda_n(k_1 + k_2)}{D^2\lambda_n^2 - k_1 k_2}. \tag{A.10}$$

The solution of the partial differential Eq. A.5 is given by the sum of the general solutions Eq. A.9 for all eigenvalues λ_n. The leading spatially inhomogeneous mode is the solution for the eigenvalue λ_1, since this function relaxes the slowest to the steady state. Hence, a good estimate for the relaxation time (τ) to the steady state is

$$\tau_{approx} = \frac{1}{D\lambda_1^2} \tag{A.11}$$

For 1000 receptor per cell, an intercellular distance of 10 µm and a diffusion coefficient $D = 36000$ µm^2/h the first eigenvalue is given by $\lambda_1 = 0.16$. For these values the relaxation time τ_{approx} is smaller than 1 second.

Appendix B

Calculating the IL-2 concentrations at the plasma membrane

After applying the quasi-steady-state approximation (Eq.2.9), the reaction-diffusion equation describing the extracellular IL-2 concentration (I, Eq.2.5) is given by

$$\frac{d^2 I}{dx^2} = \frac{k_d}{D} I. \qquad (B.1)$$

B.1 Mathematical model of the IL-2/IL-2 receptor dynamics of a single Th cell

For the mathematical model describing the IL-2/IL-2 receptor dynamics of single Th cells, it is assumed that the diffusion of IL-2 is limited by a diffusion barrier. The boundary conditions at the plasma membrane (Eq.B.2) and the diffusion barrier(Eq.B.3) are therefore

$$-D\frac{\partial I}{\partial x}\bigg|_{x=0} = q - k_{on} R_1 I + k_{off} C_1 \qquad (B.2)$$

$$-D\frac{\partial I}{\partial x}\bigg|_{x=L} = 0. \qquad (B.3)$$

Solving Equation B.1, the IL-2 distribution in the extracellular space is given by

$$I(x,t) = \frac{q + k_{off}C}{\psi\left(1 - e^{-2\sqrt{\frac{k_d}{D}}L}\right) + k_{on}R\left(1 + e^{-2\sqrt{\frac{k_d}{D}}L}\right)}\left(e^{-\sqrt{\frac{k_d}{D}}x}\right) \quad \text{(B.4)}$$

$$+ \frac{q + k_{off}C}{\psi\left(1 - e^{-2\sqrt{\frac{k_d}{D}}L}\right) + k_{on}R\left(1 + e^{-2\sqrt{\frac{k_d}{D}}L}\right)}\left(e^{-\sqrt{\frac{k_d}{D}}x - 2\sqrt{\frac{k_d}{D}}L}\right)$$

and the IL-2 concentration at the plasma membrane of the Th cell reads as

$$I(0,t) = \frac{q + k_{off}C}{\psi\chi + k_{on}R} \quad \text{(B.5)}$$

$$\psi = 10^{-24}N_AFD\sqrt{\frac{k_d}{D}} \quad \text{(B.6)}$$

$$\chi = \frac{\left(1 - e^{-2\sqrt{\frac{k_d}{D}}L}\right)}{\left(1 + e^{-2\sqrt{\frac{k_d}{D}}L}\right)}. \quad \text{(B.7)}$$

Therefore, the IL-2 concentration at the plasma membrane is given by the ratio of the IL-2 releasing reactions and the IL-2 binding to the IL-2 receptor together with a term describing the IL-2 dilution in space.

B.2   Mathematical model of two interacting T helper lymphocytes

The mathematical model describing the IL-2 receptor dynamics of two Th cells and the IL-2 diffusion in the intercellular space enables the analysis of autocrine or paracrine interaction. The Equations for the IL-2 receptor dynamics are given in Figure 2.2. Solving Equation B.1 using the following boundary conditions

$$-D\frac{\partial I}{\partial x}\bigg|_{x=0} = q_1 - k_{on}R_1I + k_{off}C_1 \quad \text{(B.8)}$$

$$-D\frac{\partial I}{\partial x}\bigg|_{x=L} = -q_2 + k_{on}R_2I - k_{off}C_2, \quad \text{(B.9)}$$

the IL-2 concentration at the plasma membrane of either Th cell is given by

$$
\begin{aligned}
I(0,t) &= \frac{\psi\left(q_1 + k_{off}C_1\right) + \psi\frac{2e^{\sqrt{\frac{k_d}{D}}L}}{1+e^{2\sqrt{\frac{k_d}{D}}L}}\left(q_2 + k_{off}C_2\right)}{\chi k_{on}^2 R_1 R_2 + \psi k_{on}(R_1 + R_2) + \chi\psi^2} \\
&+ \frac{\chi k_{on}R_2(q_1 + k_{off}C_1}{\chi k_{on}^2 R_1 R_2 + \psi k_{on}(R_1 + R_2) + \chi\psi^2}
\end{aligned}
\tag{B.10}
$$

$$
\begin{aligned}
I(L,t) &= \frac{\psi\frac{2e^{\sqrt{\frac{k_d}{D}}L}}{1+e^{2\sqrt{\frac{k_d}{D}}L}}\left(q_1 + k_{off}C_1\right) + \psi\left(q_2 + k_{off}C_2\right)}{\chi k_{on}^2 R_1 R_2 + \psi k_{on}(R_1 + R_2) + \chi\psi^2} \\
&+ \frac{\chi k_{on}R_1(q_2 + k_{off}C_2)}{\chi k_{on}^2 R_1 R_2 + \psi k_{on}(R_1 + R_2) + \chi}
\end{aligned}
\tag{B.11}
$$

$$
\chi = \frac{e^{2\sqrt{\frac{k_d}{D}}L} - 1}{e^{2\sqrt{\frac{k_d}{D}}L} + 1}
\tag{B.12}
$$

$$
\psi = N_A F D \sqrt{\frac{k_d}{D}}.
\tag{B.13}
$$

From these equations one can derive the IL-2 concentration of at the plasma membrane of a coupled Th and Treg cell pair by setting $q_2 = 0$ (cf. Equation 2.10 and 2.14).

B.3 Mathematical model of T helper cell and regulatory T cell interaction at a ration of 2:1

Co-culture assays show that the suppression of T cell proliferation is reduced but not abolished when the ratio of Th cells to Treg is increased from 1:1 to 2:1 and further [17]. To investigate if the reduction of the suppressive activity of the Treg cell can be explained by a reduction of competition, we propose the following simplified model (Figure 4.6). The important feature of the model that distinguishes it from the 1:1 model is that it 'reserves' the portion of IL-2 secreted by the Th cell away from the Treg cell for autocrine re-uptake. Thus, the Treg cell can never take up the entire secreted IL-2. The question is whether partial IL-2 deprivation of the Th cell can be sufficient to inhibit the positive feedback loop of IL-2R α-chain induction. Specifically, we assume that the Th cells can secrete IL-2 in two directions denoted with x and y, respectively. In the intercellular space between Th cells and Treg cell (x) the secreted IL-2 can be taken up by the Th cell or the Treg cell, whereas no competition takes place in the (limited) intercellular volume not facing the regulatory T cell (y). To simplify the analysis, we assume zero-flux boundary conditions at the left-hand side of the y-space. This could be due to for example another cell without IL-2 receptors. The analysis of this system can be reduced owing to the axis of symmetry indicated in Fig. 4.6. The IL-2 receptors of the Th cell now sense different environments with IL-2 receptors facing the Treg cell (R_1, C_1) and IL-2 receptors at the opposite side (R_0, C_0). Newly synthesized and recycled IL-2 receptors are shared equally between both sides of the plasma membrane. The kinetic equations for the IL-2 receptor dynamics of the Th cells thus read:

$$\frac{dR_{0,1}}{dt} \;=\; \frac{1}{2}v_1 - k_{iR}R_{0,1} - k_{on}R_{0,1}I + k_{off}C_{0,1} + \frac{1}{2}k_{rec}E_1 \quad \text{(B.14)}$$

$$\frac{dC_{0,1}}{dt} \;=\; k_{on}R_{0,1}I - k_{off}C_{0,1} - k_{iC}C_{0,1} \quad\quad\quad\quad\quad \text{(B.15)}$$

$$\frac{dE_1}{dt} \;=\; k_{iC}C_{0,1} - (k_{deg} + k_{rec})E_1 \quad\quad\quad\quad\quad \text{(B.16)}$$

Furthermore, it is assumed that the Th cell integrates the signals coming from the IL-2 receptor on both sides of its plasma membrane. The IL-2 receptor expression rate of the Th cell (v_1) depends therefore on the total number of occupied IL-2 receptors ($C_0 + C_1$).

$$v_1 = v_{10} + v_{11}\frac{(C_0 + C_1)^m}{(C_0 + C_1)^m + K^m} \quad\quad\quad\quad \text{(B.17)}$$

The dynamics of the regulatory T cell remains unchanged, only that we now explicitly consider the IL-2 receptor to be divided between two Th cells-Treg cell interfaces:

$$\frac{dR_2}{dt} = \frac{1}{2}v_2 - k_{iR}R_2 - k_{on}R_2I + k_{off}C_2 + \frac{1}{2}k_{rec}E_2 \qquad \text{(B.18)}$$

$$\frac{dC_2}{dt} = k_{on}R_2I - k_{off}C_2 - k_{iC}C_2 \qquad \text{(B.19)}$$

$$\frac{dE_2}{dt} = k_{iC}C_2 - (k_{deg} + k_{rec})E_2 \qquad \text{(B.20)}$$

Since the IL-2 receptor dynamics of the regulatory T cell describes only one half of the regulatory T cell due to the axis of symmetry indicated in Figure 4.6 the IL-2 receptor expression reads now

$$v_2 = v_{20} + v_{21}\frac{(2C_2)^m}{(2C_2)^m + K^m}. \qquad \text{(B.21)}$$

The rate constants are as given in Table 2.1. The IL-2 dynamics $(I(x,t))$ in the intracellular space between Th cell and regulatory T cell and the boundary boundary fluxes are given by

$$\frac{\partial I(x,t)}{\partial t} = D\frac{\partial^2 I}{\partial x^2} - k_dI \qquad \text{(B.22)}$$

$$J_{x,1} = -D\frac{\partial I}{\partial x}\bigg|_{x=0} = \frac{1}{2}q - k_{on}R_1I + k_{off}C_1 \qquad \text{(B.23)}$$

$$J_{x,2} = -D\frac{\partial I}{\partial x}\bigg|_{x=L_x} = k_{on}R_{12}I - k_{off}C_2 \qquad \text{(B.24)}$$

with the IL-2 secretion rate which feedback dependent term depends on the number of IL-2/IL-2 receptor complexes on the side of the plasma membrane facing the regulatory T cell and the ones which does not

$$q = q_0 + q_1\frac{(C_0 + C_1)^m}{(C_0 + C_1)^m + K_q^m}. \qquad \text{(B.25)}$$

The extracellular IL-2 concentration $(I(y,t))$ on the opposite side of the Th cell obeys:

$$\frac{\partial I(xy, t)}{\partial t} \;=\; D\frac{\partial^2 I}{\partial y^2} - k_d I \tag{B.26}$$

$$J_{y,1} \;=\; -D\frac{\partial I}{\partial x}\bigg|_{y=0} = \frac{1}{2}q - k_{on}R_0 I + k_{off}C_0 \tag{B.27}$$

$$J_{y,2} \;=\; -D\frac{\partial I}{\partial x}\bigg|_{y=L_y} = 0 \tag{B.28}$$

Exploiting again that the IL-2 diffusion is much faster than the IL-2 receptor dynamics, application of the quasi-steady-state approximation for the diffusion part reduces the problem to solving ordinary differential equations that are again readily subjected to numerical bifurcation analysis.

Appendix C

Strong coupling between IL-2 and IL-2 receptor expression

In this section of the appendix, additional analysis of the IL-2/IL-2 receptor gene-expression network is presented. The IL-2 receptor kinetics and the numbers of IL-2 receptors in the steady state are shown for a small intercellular distance of 10μm. Furthermore, the question about the transient nature of IL-2 expression is addressed. The sensitivity of the activation threshold in dependence of the parameters describing the IL-2 receptor dynamics and the IL-2 signaling is shown. At the end additional material for the interaction of two Th cells is given.

The expression of IL-2 is transient as it is described for most cytokines. The expression of IL-2 is initiated after a T cell is stimulated with its cognate antigen through the T cell receptor. The strength of the antigen stimulus controls the probability with which IL-2 is expressed as well as the amount of transcribed IL-2 [60]. After approximately 20h of stimulation the maximum of IL-2 secreting cells is reached (Figure 1.1). First the extracellular IL-2 distribution is analyzed. A situation was chosen with moderate IL-2 secretion rate and practically unlimited diffusion compared to distances occur *in vivo* (($L = 1000$ μm, cf. Figure 1.4). The extracellular IL-2 is high through out the extracellular space considered if the IL-2 receptor expression is low (Figure C.1A). After approximately 12 hours of the antigen stimulation the IL-2 receptor expression is upregulated. The extracellular IL-2 distribution drops significantly, when the IL-2 receptors are upregulated, indicating that the upregulation is due to the activation of the autocrine positive feedback loop. High IL-2 receptor expression efficiently limits the region with high IL-2 concentration to a distance of 20 μm from the plasma membrane of the Th cell. Due to its efficient re-uptake of IL-2, the upregulation of the

IL-2 receptors causes a rapid drop in IL-2 concentration, although the IL-2 secretion is assumed to be constant. This is in agreement with the fact that almost all IL-2 receptors are in complex with IL-2. (Figure C.1A). This efficient re-uptake of IL-2 results in high concentration gradients of extracellular IL-2; high IL-2 concentration in the order of nM occurs only localized at the plasma membrane and practically no IL-2 is detected at a distance of 200 µm. This finding is in agreement with experimental measurements of very low IL-2 concentration in the supernatant of cultured Th cells. These measurements reveal a very low IL-2 concentration of a few pM [61]. Which has been unexpected since the IL-2 secretion rate has been determined to be high [46]. Our simulation suggests that the high uptake rate of its receptor localizes the IL-2 concentration efficiently resulting in these measured low concentration in the bulk phase.

However, negative feedback regulation is described for IL-2 expression, which could also lead to the low IL-2 concentration observed. To illustrate that IL-2 uptake not negative feedback is responsible for the low extracellular IL-2 concentration, the consequences of transient IL-2 expression on the IL-2 receptor expression is analyzed. To determine the effect of the transient nature of IL-2 expression the IL-2 secretion rate (q) was multiplied by an appropriate function to match the IL-2 downregulation and the absence of any IL-2 secreting cell after approximately 72h (cf. Figure 1.1). In addition, a more physiological situation is considered; the IL-2 and the IL-2 receptor expression is simulated for an extracellular space of 10 µm (Figure C.1B). Again, the upregulation of the IL-2 receptor results in a rapid drop of IL-2 concentration due to high re-uptake of IL-2, subsequently downregulation of the IL-2 secretion has no influence of the extracellular IL-2 concentration anymore. In contrast, the kinetics of IL-2 receptor complexes strongly correlates with the IL-2 secretion kinetics. The downregulation of the IL-2 receptor follows the downregulation of IL-2 and is only delayed by the recycling of the receptor (Figure C.1B). Therefore, the high affinity of IL-2 to its receptor and the activation induced IL-2 receptor not only localizes the extracellular IL-2 concentration, but also induces a strong dependence on high IL-2 receptor expression to prolonged high extracellular IL-2 supply.

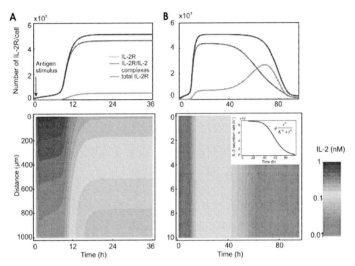

Figure C.1. Dynamics of the IL-2/IL-2 receptor expression network
(A) The upregulation of IL-2 receptor expression occurs after a delay of 12 hours
of antigen stimulation ($q_0 = 9000$ molecules/h). It is accompanied by an intense
reduction of the extracellular IL-2 concentration. Due to the strong re-uptake
of IL-2 by its receptor the extracellular IL-2 concentration is strongly localized
($L = 1000\mu m$, $k_d = 10h^{-1}$). (B) The IL-2 secretion is assumed to be transient
($q_{trans} = q * \frac{t^n}{t^n + K^n}$, $K = 60h$, $n = 6$, white box). The IL-2 receptor kinetics
follows the downregulation of the IL-2 only delayed by the recycling of the receptor
(cf. white box with upper panel). Again strong re-uptake utilizes extracellular IL-
2 almost completely also at small distances ($q_0 = 9000$ molecules/h, $L = 10\mu m$,
$k_d = 10h^{-1}$).

Appendix D

Calculating the critical distance for competition

The distance at which Th cells and Treg cells take up equal amounts of IL-2 provides an estimate for the spatial range over which the Treg cells can compete efficiently for IL-2. This distance ($L_{1/2}$) can be calculated from the ratio ρ of IL-2 uptake rate by the Th cells and the IL-2 secretion rate (q) neglecting the small degradation rate of IL-2 in the extracellular medium and the slow dissociation of IL-2 from its receptor, we have $\rho = \frac{k_{on} R_1 I_1}{q}$ (Eq. 2.10). The IL-2 concentration on the plasma membrane of the Th cell is

$$I(0,t) \quad = \quad \frac{q(1 + \frac{k_{on}}{DN_A F} LR_2)}{k_{on}(R_1 + R_2 + \frac{k_{on}}{DN_A F} LR_1 R_2)}. \tag{D.1}$$

The secreted IL-2 is shared equally between the Th cell and the regulatory T cell if $\rho = 1/2$. Solving this equation for $L_{1/2}$ yields the critical distance for cytokine competition

$$L_{1/2} = \frac{DN_A F}{k_{on}} \left(\frac{1}{R_1} - \frac{1}{R_2} \right). \tag{D.2}$$

Using the cell surface densities of receptors, r_1 and r_2 , we get:

$$L_{1/2} = \frac{D}{k_{on}} \left(\frac{1}{r_1} - \frac{1}{r_2} \right) \tag{D.3}$$

Appendix E

Analysis of Th cell proliferation

E.1 The Smith-Martin Model

The Smith-Martin Model describes the cell cycle. It divides the one round of division into a A-phase, which correspond to G1 phase, and a B-phase, which describes the S to M Phase of the cell cycle. Cell stay in the B-phase for the period of time Δ. The Smith-Martin Model is given by the following equations

$$\frac{dA_n(t)}{dt} = 2b_{n-1}(t, \Delta) - (\lambda + d_A)A_n(t) \tag{E.1}$$

$$\frac{\partial b_n(t, s)}{\partial t} + \frac{\partial b_n(t,s)}{\partial s} = -d_B b_n(t, s) \tag{E.2}$$

with $0 < s < \Delta$ and $b_n(t, 0) = \lambda A_n(t)$ [59]. The number of divisions has undergone by time t is n. $A_n(t)$ is the number of cells in the A-state and $b_n(t, s)$ the fraction of cells that are in the B-phase at the tim $t - s$. The time that the cells spend in the A-phase is exponentially distributed with the parameter λ. The duration time of the cells in the B-phase is Δ. The total number of cells in the B-phase is given by $B_n = \int_o^\Delta b_n(t, s)ds$. The cells die in the A-phase with the rate constant d_A and in the B-phase with the rate constant d_B. The Smith-Martin Model can be reduced by expressing $b_n(t, s)$ in terms of $A_n(t)$ as $b_n(s, t) = \lambda e^{-d_B s}A_n(t - s), 0 < s < 0$. Then, the

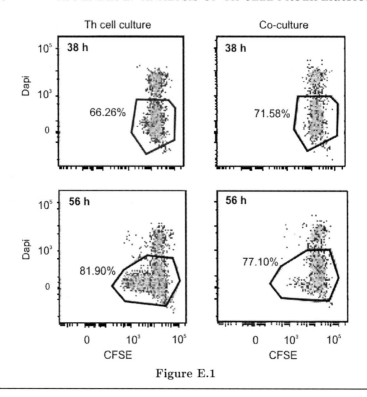

Figure E.1

Smith-Martin Model reads as follows

$$\frac{dA_n}{dt} = 2\lambda e^{-d_B\Delta}A_{n-1}(t-\Delta) - (\lambda + d_A)A_n \qquad (E.3)$$

$$B_n = \lambda \int_0^\Delta e^{-d_B s}A_n(t-s)ds. \qquad (E.4)$$

For the analysis of the Th cell proliferation in the presence of Treg cells, the state (A_0 and B_0) until the cells enter the cell cycle is considered separately. In addition no differences in the rate constants describing the apoptosis is assumed ($d_A = d_B = d$).

Bibliography

[1] Lichtman AH Abbas AK. *Cellular and Molecular Immunology 5th Edition*. Saunders, Philadelphia, 2003.

[2] Afonso R M Almeida, Bruno Zaragoza, and Antonio A Freitas. Indexation as a novel mechanism of lymphocyte homeostasis: the number of CD4+CD25+ regulatory T cells is indexed to the number of IL-2-producing cells. *J Immunol*, 177(1):192–200, Jul 2006.

[3] David Angeli, James E Ferrell, and Eduardo D Sontag. Detection of multistability, bifurcations, and hysteresis in a large class of biological positive-feedback systems. *Proc Natl Acad Sci U S A*, 101(7):1822–1827, Feb 2004.

[4] R. Antia, S. S. Pilyugin, and R. Ahmed. Models of immune memory: on the role of cross-reactive stimulation, competition, and homeostasis in maintaining immune memory. *Proc Natl Acad Sci U S A*, 95(25):14926–14931, Dec 1998.

[5] M. Assenmacher, M. Löhning, A. Scheffold, R. A. Manz, J. Schmitz, and A. Radbruch. Sequential production of IL-2, IFN-gamma and IL-10 by individual staphylococcal enterotoxin B-activated T helper lymphocytes. *Eur J Immunol*, 28(5):1534–1543, May 1998.

[6] Thomas Barthlott, Halima Moncrieffe, Marc Veldhoen, Christopher J Atkins, Jillian Christensen, Anne O'Garra, and Brigitta Stockinger. CD25+ CD4+ T cells compete with naive CD4+ T cells for IL-2 and exploit it for the induction of IL-10 production. *Int Immunol*, 17(3):279–288, Mar 2005.

[7] Yasmine Belkaid and Barry T Rouse. Natural regulatory T cells in infectious disease. *Nat Immunol*, 6(4):353–360, Apr 2005.

[8] Estelle Bettelli, Maryam Dastrange, and Mohamed Oukka. Foxp3 inter-
    acts with nuclear factor of activated T cells and NF-kappa B to repress
    cytokine gene expression and effector functions of T helper cells. *Proc
    Natl Acad Sci U S A*, 102(14):5138–5143, Apr 2005.

[9] J. A. Borghans, L. S. Taams, M. H. Wauben, and R. J. de Boer. Com-
    petition for antigenic sites during T cell proliferation: a mathematical
    interpretation of in vitro data. *Proc Natl Acad Sci U S A*, 96(19):10782–
    10787, Sep 1999.

[10] Onn Brandman, James E Ferrell, Rong Li, and Tobias Meyer. Inter-
    linked fast and slow positive feedback loops drive reliable cell decisions.
    Science, 310(5747):496–498, Oct 2005.

[11] Matthew A Burchill, Jianying Yang, Christine Vogtenhuber, Bruce R
    Blazar, and Michael A Farrar. IL-2 receptor beta-dependent STAT5
    activation is required for the development of Foxp3+ regulatory T cells.
    J Immunol, 178(1):280–290, Jan 2007.

[12] S. J. Cain and P. C. Chau. Transition probability cell cycle model. Part
    I–Balanced growth. *J Theor Biol*, 185(1):55–67, Mar 1997.

[13] S. J. Cain and P. C. Chau. Transition probability cell cycle model. Part
    II–Non-balanced growth. *J Theor Biol*, 185(1):69–79, Mar 1997.

[14] Lothar Cossel. Elektronenmikroskopischer Beitrag zur Frage der Or-
    ganisation des lymphatischen Gewebes. *Zeitschrift für Zellforschung*,
    65:199–205, 1965.

[15] Javier Cote-Sierra, Gilles Foucras, Liying Guo, Lynda Chiodetti,
    Howard A Young, Jane Hu-Li, Jinfang Zhu, and William E Paul. Inter-
    leukin 2 plays a central role in Th2 differentiation. *Proc Natl Acad Sci
    U S A*, 101(11):3880–3885, Mar 2004.

[16] Jayajit Das, Mary Ho, Julie Zikherman, Christopher Govern, Ming
    Yang, Arthur Weiss, Arup K Chakraborty, and Jeroen P Roose. Digi-
    tal signaling and hysteresis characterize ras activation in lymphoid cells.
    Cell, 136(2):337–351, Jan 2009.

[17] Maurus de la Rosa, Sascha Rutz, Heike Dorninger, and Alexander Schef-
    fold. Interleukin-2 is essential for CD4+CD25+ regulatory T cell func-
    tion. *Eur J Immunol*, 34(9):2480–2488, Sep 2004.

[18] Hans Dooms, Estelle Kahn, Birgit Knoechel, and Abul K Abbas. IL-2 induces a competitive survival advantage in T lymphocytes. *J Immunol*, 172(10):5973–5979, May 2004.

[19] Heike Dorninger. Master's thesis, Deutsche Rheumaforschungszentrum; Berlin, 2004.

[20] Warren N D'Souza and Leo Lefrançois. IL-2 is not required for the initiation of CD8 T cell cycling but sustains expansion. *J Immunol*, 171(11):5727–5735, Dec 2003.

[21] Warren N D'Souza, Kimberly S Schluns, David Masopust, and Leo Lefrançois. Essential role for IL-2 in the regulation of antiviral extra-lymphoid CD8 T cell responses. *J Immunol*, 168(11):5566–5572, Jun 2002.

[22] V. Duprez, V. Cornet, and A. Dautry-Varsat. Down-regulation of high affinity interleukin 2 receptors in a human tumor T cell line. Interleukin 2 increases the rate of surface receptor decay. *J Biol Chem*, 263(26):12860–12865, Sep 1988.

[23] V. Duprez and A. Dautry-Varsat. Receptor-mediated endocytosis of interleukin 2 in a human tumor T cell line. Degradation of interleukin 2 and evidence for the absence of recycling of interleukin receptors. *J Biol Chem*, 261(33):15450–15454, Nov 1986.

[24] V. Duprez, M. Smoljanovic, M. Lieb, and A. Dautry-Varsat. Trafficking of interleukin 2 and transferrin in endosomal fractions of T lymphocytes. *J Cell Sci*, 107 (Pt 5):1289–1295, May 1994.

[25] Christine T Duthoit, Divya J Mekala, Rajshekkhar S Alli, and Terrence L Geiger. Uncoupling of IL-2 signaling from cell cycle progression in naive CD4+ T cells by regulatory CD4+CD25+ T lymphocytes. *J Immunol*, 174(1):155–163, Jan 2005.

[26] J. S. Economou and H. S. Shin. Lymphocyte-activating factor. I. Generation and physicochemical characterization. *J Immunol*, 121(4):1446–1452, Oct 1978.

[27] Jonathan M Ellery and Peter J Nicholls. Alternate signalling pathways from the interleukin-2 receptor. *Cytokine Growth Factor Rev*, 13(1):27–40, Feb 2002.

[28] Stanley F.Farlow. *Partial Differential equations for Scientists and Engineers*. Dover, 1993.

[29] Jason D Fontenot, Jeffrey P Rasmussen, Marc A Gavin, and Alexander Y Rudensky. A function for interleukin 2 in Foxp3-expressing regulatory T cells. *Nat Immunol*, 6(11):1142–1151, Nov 2005.

[30] Gláucia C Furtado, Maria A Curotto de Lafaille, Nino Kutchukhidze, and Juan J Lafaille. Interleukin 2 signaling is required for CD4(+) regulatory T cell function. *J Exp Med*, 196(6):851–857, Sep 2002.

[31] Sarah L Gaffen and Kathleen D Liu. Overview of interleukin-2 function, production and clinical applications. *Cytokine*, 28(3):109–123, Nov 2004.

[32] R. N. Ghosh, D. L. Gelman, and F. R. Maxfield. Quantification of low density lipoprotein and transferrin endocytic sorting HEp2 cells using confocal microscopy. *J Cell Sci*, 107 (Pt 8):2177–2189, Aug 1994.

[33] F. Granucci, C. Vizzardelli, N. Pavelka, S. Feau, M. Persico, E. Virzi, M. Rescigno, G. Moro, and P. Ricciardi-Castagnoli. Inducible IL-2 production by dendritic cells revealed by global gene expression analysis. *Nat Immunol*, 2(9):882–888, Sep 2001.

[34] Thomas Höfer, Gwendolin Muehlinghaus, Katrin Moser, Taketoshi Yoshida, Henrik E Mei, Katrin Hebel, Anja Hauser, Bimba Hoyer, Elke O Luger, Thomas Dörner, Rudolf A Manz, Falk Hiepe, and Andreas Radbruch. Adaptation of humoral memory. *Immunol Rev*, 211:295–302, Jun 2006.

[35] Thomas Höfer, Holger Nathansen, Max Löhning, Andreas Radbruch, and Reinhart Heinrich. GATA-3 transcriptional imprinting in Th2 lymphocytes: a mathematical model. *Proc Natl Acad Sci U S A*, 99(14):9364–9368, Jul 2002.

[36] A. Hémar, A. Subtil, M. Lieb, E. Morelon, R. Hellio, and A. Dautry-Varsat. Endocytosis of interleukin 2 receptors in human T lymphocytes: distinct intracellular localization and fate of the receptor alpha, beta, and gamma chains. *J Cell Biol*, 129(1):55–64, Apr 1995.

[37] Patrick G Hogan, Lin Chen, Julie Nardone, and Anjana Rao. Transcriptional regulation by calcium, calcineurin, and NFAT. *Genes Dev*, 17(18):2205–2232, Sep 2003.

[38] Shohei Hori, Takashi Nomura, and Shimon Sakaguchi. Control of regulatory T cell development by the transcription factor Foxp3. *Science*, 299(5609):1057–1061, Feb 2003.

[39] Katrina K Hoyer, Hans Dooms, Luke Barron, and Abul K Abbas. Interleukin-2 in the development and control of inflammatory disease. *Immunol Rev*, 226:19–28, Dec 2008.

[40] Millie Hughes-Fulford, Eiko Sugano, Thomas Schopper, Chai-Fei Li, J. B. Boonyaratanakornkit, and Augusto Cogoli. Early immune response and regulation of IL-2 receptor subunits. *Cell Signal*, 17(9):1111–1124, Sep 2005.

[41] J. Jain, C. Loh, and A. Rao. Transcriptional regulation of the IL-2 gene. *Curr Opin Immunol*, 7(3):333–342, Jun 1995.

[42] Walport M Shlomchik MJ Janeway CA, Travers P. *Immunobiology 5th edition*. Garland, Churchill Livingstone, New York, 2001.

[43] D. M. Jelley-Gibbs, N. M. Lepak, M. Yen, and S. L. Swain. Two distinct stages in the transition from naive CD4 T cells to effectors, early antigen-dependent and late cytokine-driven expansion and differentiation. *J Immunol*, 165(9):5017–5026, Nov 2000.

[44] H. P. Kim, J. Kelly, and W. J. Leonard. The basis for IL-2-induced IL-2 receptor alpha chain gene regulation: importance of two widely separated IL-2 response elements. *Immunity*, 15(1):159–172, Jul 2001.

[45] S. Kirkpatrick, C. D. Gelatt, and M. P. Vecchi. Optimization by Simulated Annealing. *Science*, 220(4598):671–680, May 1983.

[46] W. W. Kum, S. B. Cameron, R. W. Hung, S. Kalyan, and A. W. Chow. Temporal sequence and kinetics of proinflammatory and anti-inflammatory cytokine secretion induced by toxic shock syndrome toxin 1 in human peripheral blood mononuclear cells. *Infect Immun*, 69(12):7544–7549, Dec 2001.

[47] Ruth Y Lan, Carlos Selmi, and M. Eric Gershwin. The regulatory, inflammatory, and T cell programming roles of interleukin-2 (IL-2). *J Autoimmun*, 31(1):7–12, Aug 2008.

[48] I. H. Lee, W. P. Li, K. B. Hisert, and L. B. Ivashkiv. Inhibition of interleukin 2 signaling and signal transducer and activator of transcription (STAT)5 activation during T cell receptor-mediated feedback inhibition of T cell expansion. *J Exp Med*, 190(9):1263–1274, Nov 1999.

[49] D. T. Leung, S. Morefield, and D. M. Willerford. Regulation of lymphoid homeostasis by IL-2 receptor signals in vivo. *J Immunol*, 164(7):3527–3534, Apr 2000.

[50] Abbt M Limpert E, Stahel WA. Log-normal Distributions across the Sciences: Keys and Clues. *BioScience*, 51(5):341–352, 2001.

[51] J. X. Lin and W. J. Leonard. Signaling from the IL-2 receptor to the nucleus. *Cytokine Growth Factor Rev*, 8(4):313–332, Dec 1997.

[52] Roberto A Maldonado, Darrell J Irvine, Robert Schreiber, and Laurie H Glimcher. A role for the immunological synapse in lineage commitment of CD4 lymphocytes. *Nature*, 431(7008):527–532, Sep 2004.

[53] Thomas R Malek and Allison L Bayer. Tolerance, not immunity, crucially depends on IL-2. *Nat Rev Immunol*, 4(9):665–674, Sep 2004.

[54] Athanasius F M Marée, Pere Santamaria, and Leah Edelstein-Keshet. Modeling competition among autoreactive CD8+ T cells in autoimmune diabetes: implications for antigen-specific therapy. *Int Immunol*, 18(7):1067–1077, Jul 2006.

[55] J. W. Mier and R. C. Gallo. Purification and some characteristics of human T-cell growth factor from phytohemagglutinin-stimulated lymphocyte-conditioned media. *Proc Natl Acad Sci U S A*, 77(10):6134–6138, Oct 1980.

[56] D. A. Morgan, F. W. Ruscetti, and R. Gallo. Selective in vitro growth of T lymphocytes from normal human bone marrows. *Science*, 193(4257):1007–1008, Sep 1976.

[57] H. Nakajima, X. W. Liu, A. Wynshaw-Boris, L. A. Rosenthal, K. Imada, D. S. Finbloom, L. Hennighausen, and W. J. Leonard. An indirect effect of Stat5a in IL-2-induced proliferation: a critical role for Stat5a in IL-2-mediated IL-2 receptor alpha chain induction. *Immunity*, 7(5):691–701, Nov 1997.

[58] Pushpa Pandiyan, Lixin Zheng, Satoru Ishihara, Jennifer Reed, and Michael J Lenardo. CD4+CD25+Foxp3+ regulatory T cells induce cytokine deprivation-mediated apoptosis of effector CD4+ T cells. *Nat Immunol*, 8(12):1353–1362, Dec 2007.

[59] Sergei S Pilyugin, Vitaly V Ganusov, Kaja Murali-Krishna, Rafi Ahmed, and Rustom Antia. The rescaling method for quantifying the turnover of cell populations. *J Theor Biol*, 225(2):275–283, Nov 2003.

[60] Miriam Podtschaske, Uwe Benary, Sandra Zwinger, Thomas Höfer, Andreas Radbruch, and Ria Baumgrass. Digital NFATc2 activation per cell transforms graded T cell receptor activation into an all-or-none IL-2 expression. *PLoS ONE*, 2(9):e935, 2007.

[61] Uma Prabhakar, Edward Eirikis, Manjula Reddy, Eva Silvestro, Susan Spitz, Charles Pendley, Hugh M Davis, and Bruce E Miller. Validation and comparative analysis of a multiplexed assay for the simultaneous quantitative measurement of Th1/Th2 cytokines in human serum and human peripheral blood mononuclear cell culture supernatants. *J Immunol Methods*, 291(1-2):27–38, Aug 2004.

[62] A. Raue, C. Kreutz, T. Maiwald, J. Bachmann, M. Schilling, U. Klingmüller, and J. Timmer. Structural and practical identifiability analysis of partially observed dynamical models by exploiting the profile likelihood. *Bioinformatics*, 25(15):1923–1929, Aug 2009.

[63] Y. Refaeli, L. Van Parijs, C. A. London, J. Tschopp, and A. K. Abbas. Biochemical mechanisms of IL-2-regulated Fas-mediated T cell apoptosis. *Immunity*, 8(5):615–623, May 1998.

[64] C. Rusterholz, P. C. Henrioud, and M. Nabholz. Interleukin-2 (IL-2) regulates the accessibility of the IL-2-responsive enhancer in the IL-2 receptor alpha gene to transcription factors. *Mol Cell Biol*, 19(4):2681–2689, Apr 1999.

[65] B. Sadlack, J. Löhler, H. Schorle, G. Klebb, H. Haber, E. Sickel, R. J. Noelle, and I. Horak. Generalized autoimmune disease in interleukin-2-deficient mice is triggered by an uncontrolled activation and proliferation of CD4+ T cells. *Eur J Immunol*, 25(11):3053–3059, Nov 1995.

[66] B. Sadlack, H. Merz, H. Schorle, A. Schimpl, A. C. Feller, and I. Horak. Ulcerative colitis-like disease in mice with a disrupted interleukin-2 gene. *Cell*, 75(2):253–261, Oct 1993.

[67] S. Sakaguchi, N. Sakaguchi, M. Asano, M. Itoh, and M. Toda. Immunologic self-tolerance maintained by activated T cells expressing IL-2 receptor alpha-chains (CD25). Breakdown of a single mechanism of self-tolerance causes various autoimmune diseases. *J Immunol*, 155(3):1151–1164, Aug 1995.

[68] Shimon Sakaguchi, Tomoyuki Yamaguchi, Takashi Nomura, and Masahiro Ono. Regulatory T cells and immune tolerance. *Cell*, 133(5):775–787, May 2008.

[69] Carlos Salazar and Thomas Höfer. Allosteric regulation of the transcription factor NFAT1 by multiple phosphorylation sites: a mathematical analysis. *J Mol Biol*, 327(1):31–45, Mar 2003.

[70] Alexander Scheffold, Kenneth M Murphy, and Thomas Höfer. Competition for cytokines: T(reg) cells take all. *Nat Immunol*, 8(12):1285–1287, Dec 2007.

[71] Edda G Schulz, Luca Mariani, Andreas Radbruch, and Thomas Höfer. Sequential polarization and imprinting of type 1 T helper lymphocytes by interferon-gamma and interleukin-12. *Immunity*, 30(5):673–683, May 2009.

[72] Ethan M Shevach, Richard A DiPaolo, John Andersson, Dong-Mei Zhao, Geoffrey L Stephens, and Angela M Thornton. The lifestyle of naturally occurring CD4+ CD25+ Foxp3+ regulatory T cells. *Immunol Rev*, 212:60–73, Aug 2006.

[73] Stanislav Y Shvartsman, Cyrill B Muratov, and Douglas A Lauffenburger. Modeling and computational analysis of EGF receptor-mediated cell communication in Drosophila oogenesis. *Development*, 129(11):2577–2589, Jun 2002.

[74] J. A. Smith and L. Martin. Do cells cycle? *Proc Natl Acad Sci U S A*, 70(4):1263–1267, Apr 1973.

[75] K. A. Smith and D. A. Cantrell. Interleukin 2 regulates its own receptors. *Proc Natl Acad Sci U S A*, 82(3):864–868, Feb 1985.

[76] Kendall A Smith. The quantal theory of immunity. *Cell Res*, 16(1):11–19, Jan 2006.

[77] Dorothy K Sojka, Denis Bruniquel, Ronald H Schwartz, and Nevil J Singh. IL-2 secretion by CD4+ T cells in vivo is rapid, transient, and influenced by TCR-specific competition. *J Immunol*, 172(10):6136–6143, May 2004.

[78] Dorothy K Sojka, Angela Hughson, Teresa L Sukiennicki, and Deborah J Fowell. Early kinetic window of target T cell susceptibility to CD25+ regulatory T cell activity. *J Immunol*, 175(11):7274–7280, Dec 2005.

[79] H. Suzuki, T. M. Kündig, C. Furlonger, A. Wakeham, E. Timms, T. Matsuyama, R. Schmits, J. J. Simard, P. S. Ohashi, and H. Griesser. Deregulated T cell activation and autoimmunity in mice lacking interleukin-2 receptor beta. *Science*, 268(5216):1472–1476, Jun 1995.

[80] H. Suzuki, Y. W. Zhou, M. Kato, T. W. Mak, and I. Nakashima. Normal regulatory alpha/beta T cells effectively eliminate abnormally activated T cells lacking the interleukin 2 receptor beta in vivo. *J Exp Med*, 190(11):1561–1572, Dec 1999.

[81] T. Takahashi, Y. Kuniyasu, M. Toda, N. Sakaguchi, M. Itoh, M. Iwata, J. Shimizu, and S. Sakaguchi. Immunologic self-tolerance maintained by CD25+CD4+ naturally anergic and suppressive T cells: induction of autoimmune disease by breaking their anergic/suppressive state. *Int Immunol*, 10(12):1969–1980, Dec 1998.

[82] A. M. Thornton and E. M. Shevach. Suppressor effector function of CD4+CD25+ immunoregulatory T cells is antigen nonspecific. *J Immunol*, 164(1):183–190, Jan 2000.

[83] Kevin Thurley. Numerical and analytical solutions of a spatio-temporal model for interleukin-2 expression in T lymphocytes (in German). Master's thesis, Humboldt University Berlin, 76p., 2007.

[84] Dario A A Vignali, Lauren W Collison, and Creg J Workman. How regulatory T cells work. *Nat Rev Immunol*, 8(7):523–532, Jul 2008.

[85] Alejandro V Villarino, Cristina M Tato, Jason S Stumhofer, Zhengju Yao, Yongzhi K Cui, Lothar Hennighausen, John J O'Shea, and Christopher A Hunter. Helper T cell IL-2 production is limited by negative feedback and STAT-dependent cytokine signals. *J Exp Med*, 204(1):65–71, Jan 2007.

[86] H. M. Wang and K. A. Smith. The interleukin 2 receptor. Functional consequences of its bimolecular structure. *J Exp Med*, 166(4):1055–1069, Oct 1987.

[87] D. M. Willerford, J. Chen, J. A. Ferry, L. Davidson, A. Ma, and F. W. Alt. Interleukin-2 receptor alpha chain regulates the size and content of the peripheral lymphoid compartment. *Immunity*, 3(4):521–530, Oct 1995.

[88] Matthew A Williams, Aaron J Tyznik, and Michael J Bevan. Interleukin-2 signals during priming are required for secondary expansion of CD8+ memory T cells. *Nature*, 441(7095):890–893, Jun 2006.

[89] Yongqing Wu, Madhuri Borde, Vigo Heissmeyer, Markus Feuerer, Ariya D Lapan, James C Stroud, Darren L Bates, Liang Guo, Aidong Han, Steven F Ziegler, Diane Mathis, Christophe Benoist, Lin Chen,

and Anjana Rao. FOXP3 controls regulatory T cell function through cooperation with NFAT. *Cell*, 126(2):375–387, Jul 2006.

[90] Wen Xiong and James E Ferrell. A positive-feedback-based bistable 'memory module' that governs a cell fate decision. *Nature*, 426(6965):460–465, Nov 2003.

Acknowledgment

This work is dedicated to my parents, especially my father.

First I want to thank my supervisor Prof. Dr. Thomas Höfer to provide me with this interesting topic. I want to thank him for the very intensive supervision at the beginning of my PHD. At later stages, the opportunity to plan the ongoing of my project with more personal responsibility was always accompanied by his help if needed. In addition, as part of his group, I had the opportunity to cooperate closely with experimental working scientists, among I especially want to mention Prof. Dr. Max Löhning. I profited a lot from these experiences. Thus, I want to thank Prof. Höfer for this as well.

I want to thank Alexander Scheffold for sharing his experimental results without which this subject would not be able to develop. I also want to thank him for the opportunity to work in his lab. I will definitely remember the welcoming atmosphere, the patience and help he and his group provided.

I want to thank Michael Floßdorf and Edda Schulz for the help and various discussion on data fitting and statistics.

While writing this thesis the yogic advises of Sreekanth Korathiyattu were of great help. I want to thank Edda Schulz especially for dealing with the first version. Bente Kofahl, Jana Schütze and Antonio Politi for helping with the corrections. And I want to thank Edda Schulz and Luca Mariani for the nice and productive time in the last several months. And Prof. Herzel and Prof. Hammerstein for letting us being part of the ITB.

Publications

Busse, D., de la Rosa, M., Hobiger, K., Thurley, K., Floßdorf, M., Scheffold, A. and Höfer, T. (2010) Competing feedback loops shape IL-2 signaling between helper and regulatory T lymphocytes in cellular microenvironments. Proc. Natl. Acad. Sci. USA, 107, 3058-3063 .

Löhning, M., Hegazy, A.N., Pinschewer, D.D., Busse, D., Lang, K.S., Höfer, T., Radbruch, A., Zinkernagel, R.M, and Hengartner, H. (2008) Long-lived virus-reactive memory T cells generated from purified cytokine-secreting T helper type 1 and type 2 effectors. J. Exp. Med. 205, 53-61.

Löhning, M., Hegazy, A.N., Pinschewer, D.D., Busse, D., Lang, K.S., Höfer, T., Radbruch, A., Zinkernagel, R.M, and Hengartner, H. (2009) Linear differentiation of virus-reactive memory T cells generated from purified cytokine-secreting T helper type 1 and type 2 effectors. Wiener Klinische Wochenschrift 120, 62. (Meeting abstract)

Selbständigkeitserklärung

Ich versichere hiermit, die vorliegende Arbeit selbständig und ausschließlich unter Verwendung der angegebenen Mittel und ohne unerlaubte Hilfe angefertigt zu haben.

Berlin, den 17.08.2009

Dorothea Busse